Primer of

Oil and Gas Measurement

by

S. T. Horton

Published by
Petroleum Extension Service
Division of Continuing Education
The University of Texas at Austin

in cooperation with the
American Petroleum Institute's
Committee on Petroleum Measurement
(COPM)

1993

Library of Congress Cataloging-in-Publication Data

Horton, Sheryl T., 1948-
Primer of oil and gas measurement / by Sheryl T. Horton.
p. cm.
Includes index.
ISBN 0-88698-160-3
1. Petroleum—Measurement. 2. Natural gas—Measurement
I. University of Texas at Austin. Petroleum Extension Service.
II. Title.
TP691.H65 1993
665.5'028'7—dc20 93-4436
CIP

First edition published 1993, Third Impression 1994
Printed in the United States of America

Catalog No. 8.20010
ISBN 0-88698-160-3

Contents

Figures

Tables

Preface

This *Primer of Oil and Gas Measurement* introduces inexperienced personnel to the basic procedures, standards, and instruments used to measure oil and gas. It is intended as a primer for both those who measure oil and gas and those who witness measurement procedures. It can be used as an introduction for people new to the industry or as a reference for those knowledgeable about other areas of the industry but unfamiliar with measurement procedures and practices.

This book is published in cooperation with the American Petroleum Institute's Committee on Petroleum Measurement (COPM) and the Committee on Measurement Education and Training (COMET). The information presented herein is based on API standards current in the year of publication.

Acknowledgments

The knowledge and talents of many persons and the resources of many organizations were used to produce this manual. The creative support provided by the individuals involved is gratefully acknowledged. Special thanks are extended to **Jerry Upton**, Shell Oil Co., chairman of the Committee on Measurement Education and Training (COMET), for involving API in this project and making sure that any help I needed was available. Thanks also to COMET member **Jim Strawn,** The Strawn Group, for his invaluable help in reading several drafts of the manuscript, finding additional experts to review the manuscript, and researching technical matters for me. His assistance and patience are much appreciated.

In addition to these two individuals, other members of API COMET devoted many hours to reviewing the material, providing technical data, and helping to ensure the accuracy and usefulness of the book. Petex and I are grateful for their efforts. We do, however, assume full responsibility for the content as published. The following measurement specialists reviewed this manuscript:

Clarence L. Strance, Select Technology Services
Larry Pitts, Daniel Industries, Inc.
E. L. Upp, Daniel Industries, Inc.
R. E. Beaty, Amoco Production Co.
Ray Teyssandier, Texaco, Inc.
Dan Comstock, SGS Control Services, Inc.
Rinaldo V. Garrettson, Buckeye Pipeline
Robert C. Wainscott, SGS Control Services, Inc.
Robert A. Webb, Marathon Oil Co.
W. G. Poynter, Kam Controls, Inc.
Del J. Major, Texaco Trading and Transportation Inc.
Ken Embry, Perry Equipment Corp.

Without the assistance and support of several Petex staff members I could not have written this book. Special thanks to Kathy Bork for reading this manuscript more times than one person should be expected to do so and for making many valuable suggestions to improve it; to Doris Dickey for painstakingly proofreading the text; to Debbie Caples for taking the manuscript, photos, and barely legible notes and, through the art of layout, making it into a book; and to Ron Baker for giving me the opportunity to write this book and therefore work with supportive people interested in training others as best we can. Special thanks also to Jonell Clardy for her superb illustrations.

Introduction

Measurement. Custody transfer. Relative density. Words and phrases that don't necessarily prompt pictures to form in the mind's eye. Tank calibration. Sulfur content. S&W limits. API-recommended standards. More phrases that most of us are not that familiar with. Yet, for some, these are simple, easy-to-understand terms. They use them every day when buying and selling oil and oil products. These people have a particular definition in mind when they say "gauging a tank height" or "meter proving." And the definitions aren't that complicated.

Take the word *measure,* for example. To a measurement specialist it means to determine the quantity and the quality of a substance, because quantity and quality affect price. In layperson's terms, all this means is finding out how much oil or product you have and determining exactly what type of oil or product it is. Just as in any other business, you can't determine price until you know *what* you've got and how *much* of it there is.

Custody transfer may conjure up an image of a felon being transferred from county jail to state prison. In the oil business, however, it simply refers to the oil or oil product changing hands, from one vendor to another. Every time the oil or product changes hands, you need to know exactly how much and what type changed custody. If you don't know these two things you have no way of keeping track of the substance.

How, then, do you measure this oil or product? If I'm going to sell you a barrel of oil for so much per barrel, first we have to agree on the quality of that oil. Is it light or heavy? Does it have a high sulfur content? Does it contain a large amount of sediment and water—something that is of no value to either of us? Next we have to agree that, yes, the amount we're transferring *is* a barrel. In short, we must test and then measure the oil based on a standard of measurement.

API, the American Petroleum Institute, recommends certain measurement procedures, definitions, and equipment that many industry people rely on. The standards were written with the help of industry committees. A buyer and a seller could agree on standards other than API's if they wanted to, but the API standards are widely respected and used.

One very important standard involves the temperature of the oil or the product. Oil and oil products expand when warm (i.e., they take up more volume); they contract (take up less volume) when cold. The relative density of the oil or the product determines how much it expands or contracts. Density refers to a substances's ratio of mass to space. A denser substance, like heavy crude oil, will not expand or contract as much as a less-dense substance like kerosene. So, when measuring, you use tables that help convert the measured volume of a particular liquid to what the volume would be at 60°F, the standard that the industry has agreed to use.

When you are measuring natural gas, another standard involves pressure. Because gas expands and contracts so easily, its volume is directly affected by both temperature and pressure. At a constant temperature, if you double the pressure, you reduce the gas volume by half. At a constant pressure, if you double the temperature, you double the gas volume. In most environments, neither the pressure nor the temperature is constant; thus both measurements must be converted to the standard agreed-on temperature and pressure.

We will discuss other standards later in the book, but, as you can see, even though there are many things to know and understand about measuring petroleum and its products, when translated into nontechnical terms, they're not hard to understand. As a specialist once said, "I could talk measurement all day." That's what we intend to do—"talk measurement" so you can understand just what people are doing when they gauge a tank, grab a sample, calibrate a tank, or prove a meter.

CHAPTER 1

Tank Measurement

INTRODUCTION

Oil and oil products are often stored in large stationary tanks. Crude oil is stored in tanks near producing wells, at tank farms along the route to the refinery, and at the refinery itself. Oil products are stored in tanks in many areas, from the refinery to the petrochemical plant to the wholesale distributor. These liquids are stored for various reasons, but no matter what the reason, whenever they are kept in tanks, the amount stored must be accounted for and therefore must be measured precisely.

One way to measure the volume of a stored liquid is to determine the height of the liquid in the tank and then to refer to a *capacity table* computed specifically for that tank. This capacity table (also sometimes referred to as a tank table, a strapping table, or a gauge table) tells you the gross volume of liquid contained in a particular tank at any given level.[1] For example, in table 1.1 you can see that the height of the liquid is listed in one-inch increments. Opposite each height is the volume in barrels. Thus, if the liquid height in this tank is 25 feet, 6 inches, it holds 35,429.93 barrels of liquid. If the liquid height includes a fraction of an inch, you consult the end of the table for the number of barrels for each fraction of liquid height. For example, if the liquid height is 25 feet, 6¼ inches, you look at the end of the table and find that for every quarter of an inch you add 29.16 barrels to the amount listed in the rest of the table. Thus, in this example, 25 feet, 6¼ inches equals 35,459.09 barrels of liquid.

A capacity table like the one shown in table 1.1 is developed from specific measurements taken from the storage tank. At various times during its life a storage tank is carefully measured by a professional independent contractor to obtain certain dimensions, including, among others, the depth of the tank (both inside and outside), the circumference of each ring of the tank and the height of the liquid, if any, in the tank. These measurements are then forwarded to the department that computes capacity tables. Measuring a tank so that its capacity can be computed is sometimes referred to as *tank strapping*. It is also frequently called *tank calibration*.

[1] To determine the *net* volume of liquid in a tank, and thus to determine tank liquid volume accurately, you must also know the height of the liquid, the temperature and gravity of the liquid, and the sediment and water content.

TABLE 1.1
Capacity Table

XYZ REFINERY
CITY, STATE

TANK
111

TOTAL GAUGE HEIGHT—50′00″ LOCATED 2′00″ IN FROM SHELL
CAPACITIES IN BARRELS

25	34,729.88	**30**	41,730.40	**35**	48,733.23	**40**	55,737.60	**45**	62,743.29	**50**	
1	34,846.56	1	41,847.08	1	48,849.97	1	55,854.36	1	62,860.05		
2	34,963.23	2	41,963.75	2	48,966.71	2	55,971.12	2	62,976.81		
3	35,079.91	3	42,080.43	3	49,083.45	3	56,087.88	3	63,093.57		
4	35,196.58	4	42,197.10	4	49,200.19	4	56,204.65	4	63,210.34		
5	35,313.26	5	42,313.78	5	49,316.93	5	56,321.41	5	63,327.10		
6	35,429.93	6	42,430.45	6	49,433.67	6	56,438.17	6	63,443.86		
7	35,546.61	7	42,547.13	7	49,550.41	7	56,554.93	7	63,560.62		
8	35,663.29	8	42,663.80	8	49,667.15	8	56,671.69	8	63,677.38		
9	35,779.96	9	42,780.48	9	49,783.89	9	56,788.45	9	63,794.14		
10	35,896.64	10	42,897.15	10	49,900.62	10	56,905.21	10	63,910.90		
11	36,013.31	11	43,013.83	11	50,017.36	11	57,021.98	11	64,027.67		
26	36,129.99	**31**	43,130.50	**36**	50,134.10	**41**	57,138.74	**46**	64,144.43		
1	36,246.66	1	43,247.18	1	50,250.84	1	57,255.50	1			
2	36,363.34	2	43,363.85	2	50,367.58	2	57,372.26	2			
3	36,480.01	3	43,480.53	3	50,484.32	3	57,489.02	3			
4	36,596.69	4	43,597.21	4	50,601.06	4	57,605.78	4			
5	36,713.36	5	43,713.88	5	50,717.80	5	57,722.55	5			
6	36,830.04	6	43,830.56	6	50,834.54	6	57,839.31	6			
7	36,946.71	7	43,947.23	7	50,951.28	7	57,956.07	7			
8	37,063.39	8	44,063.91	8	51,068.02	8	58,072.83	8			
9	37,180.06	9	44,180.58	9	51,184.76	9	58,189.59	9			
10	37,296.74	10	44,297.26	10	51,301.50	10	58,306.35	10			
11	37,413.41	11	44,413.93	11	51,418.24	11	58,423.11	11			
27	37,530.09	**32**	44,530.61	**37**	51,534.98	**42**	58,539.88	**47**			
1	37,646.77	1	44,647.35	1	51,651.72	1	58,656.64	1			
2	37,763.44	2	44,764.09	2	51.768.46	2	58,773.40	2			
3	37,880.12	3	44,880.83	3	51,885.20	3	58,890.16	3			
4	37,996.79	4	44,997.57	4	52,001.94	4	59,006.92	4			
5	38,113.47	5	45,114.31	5	52,118.68	5	59,123.68	5			

Table 1.1—*continued*

TOTAL GAUGE HEIGHT—50′00″ LOCATED 2′00″ IN FROM SHELL CAPACITIES IN BARRELS										
6	38,230.14	6	45,231.04	6	52,234.41	6	59,240.44	6		
7	38,346.82	7	45,347.78	7	52,352.15	7	59,357.21	7		
8	38,463.49	8	45,464.52	8	52,468.89	8	59,473.97	8		
9	38,580.17	9	45,581.26	9	52,585.63	9	59,590.73	9		
10	38,696.84	10	45,698.00	10	52,702.37	10	59,707.49	10		
11	38,813.52	11	45,814.74	11	52,819.11	11	59,824.25	11		
28	38,930.19	**33**	45,931.48	**38**	52,935.85	**43**	59,941.01	**48**		
1	39,046.87	1	46,048.22	1	53,052.59	1	60,057.78	1		
2	39,163.54	2	46,164.96	2	53,169.33	2	60,174.54	2		
3	39,280.22	3	46,281.70	3	53,286.07	3	60,291.30	3		
4	39,396.89	4	46,398.44	4	53,402.81	4	60,408.06	4		
5	39,513.57	5	46,515.18	5	53,519.55	5	60,524.82	5		
6	39,630.25	6	46,631.92	6	53,636.29	6	60,641.58	6		
7	39,746.92	7	46,748.66	7	53,753.03	7	60,758.34	7	**Fractions**	
8	39,863.60	8	46,865.40	8	53,869.77	8	60,875.11	8		
9	39,980.27	9	46,982.14	9	53,986.51	9	60,991.87	9	1/16	7.29
10	40,096.95	10	47,098.88	10	54,103.25	10	61,108.63	10	1/8	14.58
11	40,213.62	11	47,215.62	11	54,219.99	11	61,225.39	11	3/16	21.87
29	40,330.30	**34**	47,332.36	**39**	54,336.73	**44**	61,342.15	**49**	1/4	29.16
1	40,446.97	1	47,449.10	1	54,453.47	1	61,458.91	1	5/16	36.45
2	40,563.65	2	47,565.83	2	54,570.20	2	61,575.67	2	3/8	43.74
3	40,680.32	3	47,682.57	3	54,686.94	3	61,692.44	3	7/16	51.03
4	40,797.00	4	47,799.31	4	54,803.68	4	61,809.20	4	1/2	58.32
5	40,913.67	5	47,916.05	5	54,920.42	5	61,925.96	5	9/16	65.62
6	41,030.35	6	48,032.79	6	55,037.16	6	62,042.72	6	5/8	72.91
7	41,147.02	7	48,149.53	7	55,153.90	7	62,159.48	7	11/16	80.20
8	41,263.70	8	48,266.27	8	55,270.64	8	62,276.24	8	3/4	87.49
9	41,380.37	9	48,383.01	9	55,387.38	9	62,393.01	9	13/16	94.78
10	41,497.05	10	48,499.75	10	55,504.12	10	62,509.77	10	7/8	102.07
11	41,613.73	11	48,616.49	11	55,620.86	11	62,626.53	11	15/16	109.36

Liquid Head Calculated at 60.0 API.
Safe Fill Height—44′00″. Overflow @ 46′00″

Strapped: June 1986
Computed: G.D.B.
File No.: 22222/B222
Calculated per API 2550
October 1965 Edition

Figure 1.1. Upright cylindrical tank with a fixed roof

To illustrate how tanks are strapped, this chapter chronicles the work done by a strapping crew on a common type of tank used in the industry—the upright cylindrical tank with a fixed roof (fig. 1.1). This type of tank is constructed of relatively wide rings of metal stacked one on top of the other and welded or bolted together. Each ring comprises several sheets of metal of uniform height and thickness welded or bolted together and fastened end-to-end to form a circle.

Tanks must be measured very precisely. A slight mistake and the volume calculations are wrong and someone may lose a lot of money. So, tanks are strapped by a crew of three or four people thoroughly trained in measurement procedures.The strapping procedures described in this chapter are also described in the API Standard 2550, *Measurement and Calibration of Upright Cylindrical Tanks,* which is part of the *Manual of Petroleum Measurement Standards* (*MPMS*), chapter 2. (For strapping procedures for other types of tanks, refer to other standards in the API *MPMS.* If the optical reference line method is used to strap a tank, the size of the crew is usually only two or three people. Use API *MPMS* chapter 2.2B in conjunction with API Standard 2550.)

PRELIMINARY TASKS

The first thing a crew does on arrival at the tank location is to check with the person in charge of the tank farm to verify which tank is to be strapped and to confirm that it is ready for strapping. To be ready, the tank must have been completely filled at least once with liquid that has the same or greater density as the liquid that it will normally hold, because the weight of the liquid puts stress on the tank shell and bottom and stabilizes the tank on its pad. Putting stress on the tank also eliminates construction distortions and helps ensure that the capacity tables are accurate. The tank should be empty and gas-free when strapped. With an empty tank the crew can measure the thickness of the tank walls and any elevations or various appurtenances like manways and overflow lines that may extend into the interior of the tank shell and alter its volume. The crew can also make sure there are no obstructions blocking the depth measurements made inside the tank and check the bottom of the tank for obstructions or distortions. Since the tank is not always empty when it is strapped, however, the crew should get the *as-built prints* of the tank from the tank farm supervisor. These prints provide the tank table department with the ring thickness, bottom, and appurtenance details that are necessary to compute tank capacity accurately.

After checking in with the person in charge, the crew locates the tank and checks that it is the correct one by finding and recording its identification number (usually stenciled on the side of the tank). Then a crew member locates the *builder's data plate* (fig. 1.2) and records the information found on it. The data plate gives the crew such information as the name of the manufacturer, the tank's erection date, and its nominal diameter, height, and capacity. All tanks built to API specifications should have a data plate. It gives the crew the basic measurements it needs to start strapping and indicates how long the strapping tapes and other measuring instruments must be. Once the tank identification number and other information are recorded, the site operator closes all inlet and outlet valves to make sure that no liquids are allowed to enter or leave the tank while it is being strapped. If the liquid level in the tank changes while it is being strapped, the stress on the tank changes and the capacity tables will not be accurate.

Figure 1.2. Builder's data plate

MEASURING TAPES

To measure the capacity of the tank, the crew must first measure the circumference of each ring, starting with the ring at the bottom of the tank. The first circumference measurement is made with a *standard* tape. Each subsequent ring is measured with a *working* tape (fig. 1.3).

A standard tape has either been calibrated by the National Institute of Standards and Technology (NIST) or calibrated against an NIST-calibrated tape. In other words, some agencies and manufacturers use a tape from NIST to calibrate another. Both types are certified to be a certain length, usually 100 feet, at a standard temperature and tension. Tapes are marked in feet or metres, with the smallest division in hundredths of a foot or in millimetres. NIST specifies the tape's length at a standard temperature (68°F) and tension because metal expands and contracts and heat and tension affect its length. The crew notes the ambient temperature when it straps the tank. The difference between the ambient temperature and 68°F is taken into consideration and factored into the capacity tables for that tank.

The crew is primarily concerned with how much tension to apply to the standard tape to get an accurate tape reading. Tension must be enough to hold the tape flat against the side of the tank and it should be the same amount of tension used when the tape was calibrated by NIST (fig. 1.4). This information is contained in a calibration report accompanying each standard tape.

A working tape has not been calibrated by NIST or any other agency or manufacturer; instead it is calibrated by the crew using the standard tape. (See "Calibrating the Working Tape"

Figure 1.3. Measuring tapes

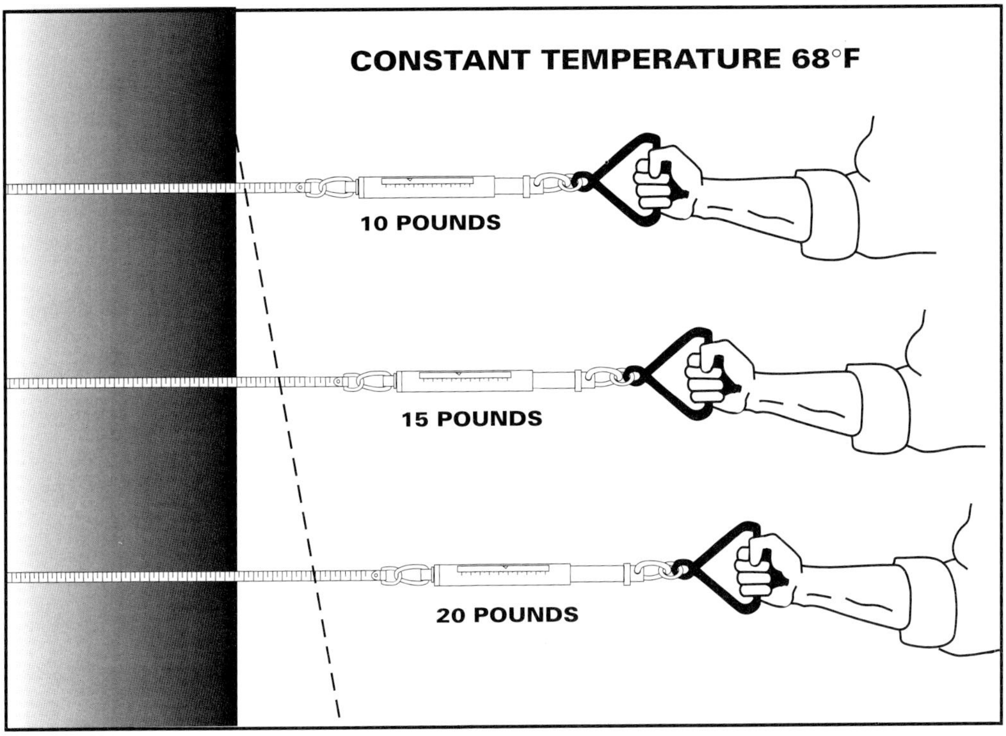

Figure 1.4. Different amounts of tension produce different readings with the same tape.

later in this chapter.) In general, the working tape is longer than the standard tape—usually long enough to circle the tank in one wrap. It is also marked in feet or metres, with the smallest division in hundredths of a foot or in millimetres. To reduce the risk of damaging the standard tape, the crew uses the working tape to measure all rings but the bottom. It is fairly easy to damage a tape when strapping a tank, and once a standard tape is damaged, it is no longer calibrated and cannot be used. Replacing the standard tape is much more expensive than replacing a working tape.

MEASURING THE FIRST RING

When a strapping crew is ready to begin measuring the bottom ring on the tank, the first thing it does is check the number embossed on the standard tape to be sure it matches the number on the calibration report. Next crew members check the bottom ring for obstructions that might block the path of the tape. All circumferences must be measured along a specific path around the ring. API recommends that this path be at a distance equal to 20% of the ring width, measured from the top of the ring (fig. 1.5). It recommends this distance because a ring bulges slightly when the tank is filled with liquid. It expands the most in the center and the least at the top and bottom. The 20% distance best represents the point for finding true volume. If there are any obstructions blocking the designated path, a substitute path as close to the 20% distance as possible is chosen. If approximating the 20% path is impossible, the crew may measure the circumference at a distance equal to 20% of the ring width measured from the bottom.

To make sure the tape stays on the correct path around the ring, the crew marks the tank shell with crayon or chalk and a *strapping pole* (fig. 1.6). This aluminum pole is adjustable so that it can be used on several rings, not just the bottom one. It has a clamp, the location of which is also adjustable, that holds the marking chalk or crayon. At the tip of the pole is a reference point used for positioning the top of the pole at the top of the ring. To mark the tape path, the crew measures the 20% or substitute path distance from the top of the pole, secures the clamp and

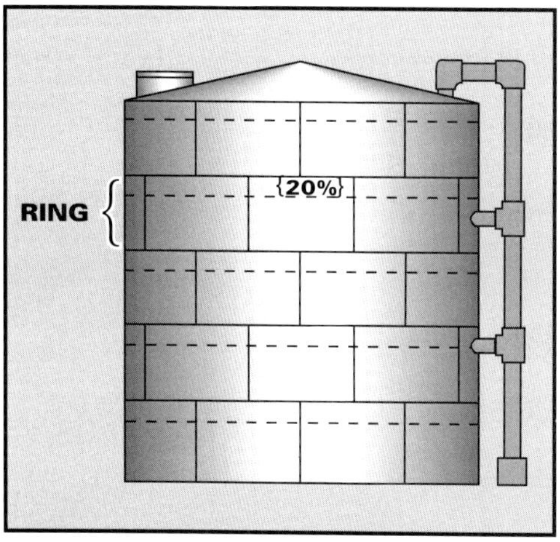

Figure 1.5. Recommended tape path

Figure 1.6. Strapping pole

chalk to the pole at that distance, and then places the strapping pole so that the reference point at the top of the pole rests on the top of the ring to be measured. While holding the pole and chalk against the tank shell, the crew circles the tank, making either a continuous chalk mark or a series of chalk dashes to designate the path of the tape (fig. 1.7).

Once the tape path is marked, a crew member takes a hammer and punch and lightly taps the tank shell to make a small indentation that establishes an anchor point for holding the standard tape. (On painted tanks, the crew must take care not to break the paint seal.) While one person anchors the tape, another makes a vertical score on the tank where the tape's zero reading falls (fig. 1.8). Then the crew extends the tape around the ring, making certain that it stays precisely on the path marked. With an instrument called a *tape clamp*

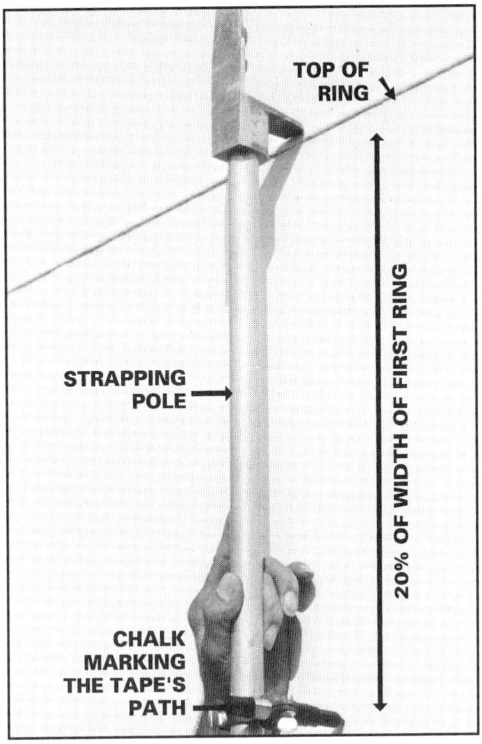

Figure 1.7. Using a strapping pole to mark tape path around first ring

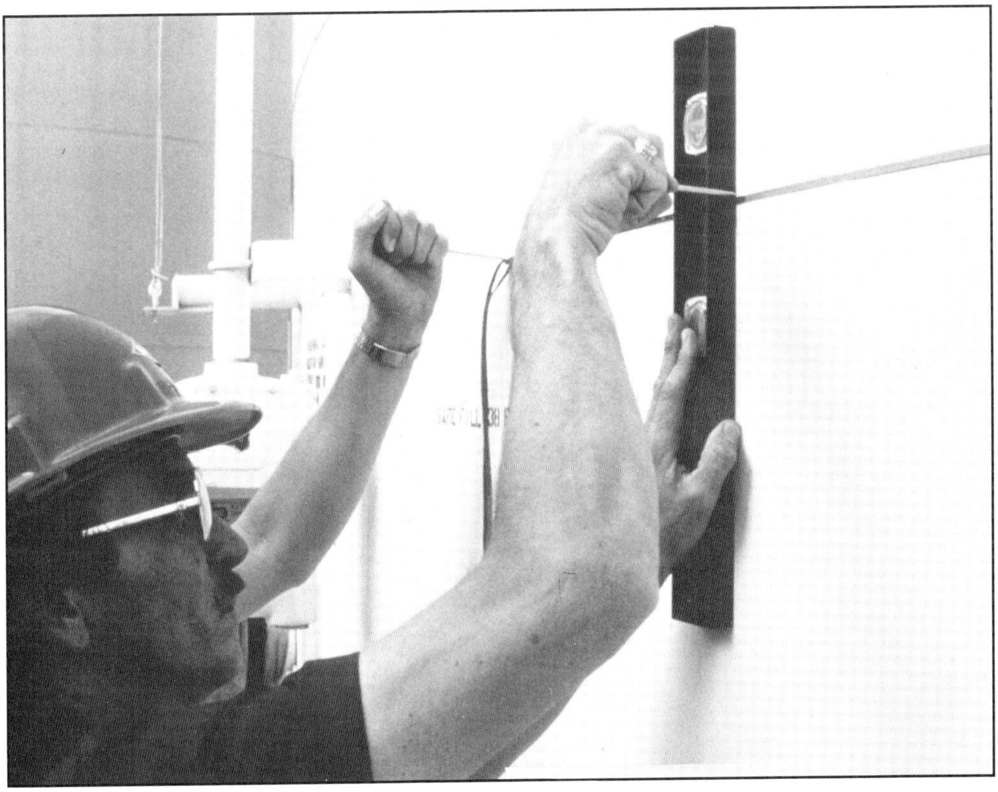

Figure 1.8. Marking the zero point to begin circumference measurement

(fig. 1.9), a crew member gently pulls until the required amount of tension has been applied to the tape. A *tension gauge* (fig. 1.9), which tells the crew how much tension is being applied, is attached to the opposite end of the tape from the clamp. The required tension is read from the calibration report accompanying the standard tape and must be at least enough to hold the tape against the tank.

Most upright tanks have a circumference of more than 100 feet, so when the 100-foot standard tape is fully extended, the crew makes a vertical score on the tank shell at the 100-foot reading. Then the crew releases tension on the tape, moves the tape forward, and lines up the zero reading on the tape with the 100-foot vertical score. Crew members center punch the shell again and anchor the standard tape with its zero reading precisely on the 100-foot vertical score. They pull on the tape until the required amount of tension is applied and extend it along the marked path. If the tank's circumference is greater than 200 feet, they again mark the tank shell, at the 200-foot mark, release tension on the tape, move the tape forward, and line up the zero reading with the 200-foot mark. They continue doing this until the tape overlaps the original zero vertical score.

At this point, the crew reads the tape and records the measure to the nearest hundredth of a foot and estimates to the nearest thousandth of a foot. That reading is added to the number of feet already measured. For example, if the tape were extended to the 100-foot mark and then extended again until it reached 89.56 feet beyond the original zero mark, 100 would be added to 89.56 to determine that the bottom ring's circumference was 189.56 feet. The crew would use this measurement to calibrate the working tape.

Figure 1.9. Tape clamp and tension gauge

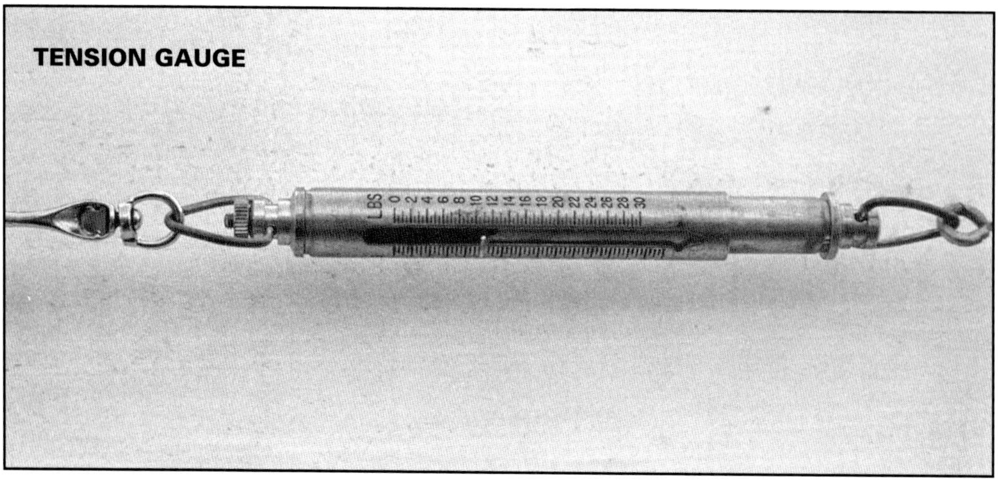

CALIBRATING THE WORKING TAPE

The crew anchors the working tape—which is generally long enough to circle the tank in one wrap—on the same path used for the standard tape. Crew members pull in the slack, checking to be sure the tape is on the chalk-marked path. They slide the working tape back and forth to break any frictional resistance that may occur between tank and shell. This is important because, when tension is applied to the tape, it must be spread evenly over the entire length. If the tape clings to the shell's surface at any point, it will be tensioned unevenly and measurement will not be accurate (fig. 1.10).

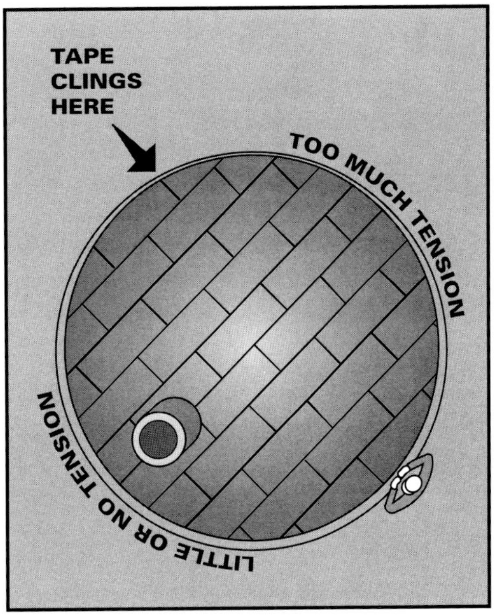

Figure 1.10. Crew must slide tape back and forth to spread tape tension evenly around tank.

After breaking the surface tension, the crew uses one of two methods to calibrate the working tape to the standard tape. In one method, the crew members apply the same amount of tension to the working tape as they applied to the standard tape. Then they record the measurement made by the working tape. If the working tape circumference equals the standard tape circumference, the tape is in calibration and no correction is needed. If, however, the working-tape circumference is *less* than the standard tape circumference, the difference between the two measurements must be added to each circumference reading taken from the working tape:

184.625	Measure on standard tape with appropriate tension
184.615	Measure on working tape with appropriate tension
000.010	Difference to be added to each 100-foot circumference reading made by working tape

If the working-tape circumference is more than the standard circumference, the difference must be subtracted from each circumference reading taken from the working tape:

184.625	Measure on standard tape with appropriate tension
184.635	Measure on working tape with appropriate tension
000.010	Difference to be subtracted from each 100-foot circumference reading made with working tape

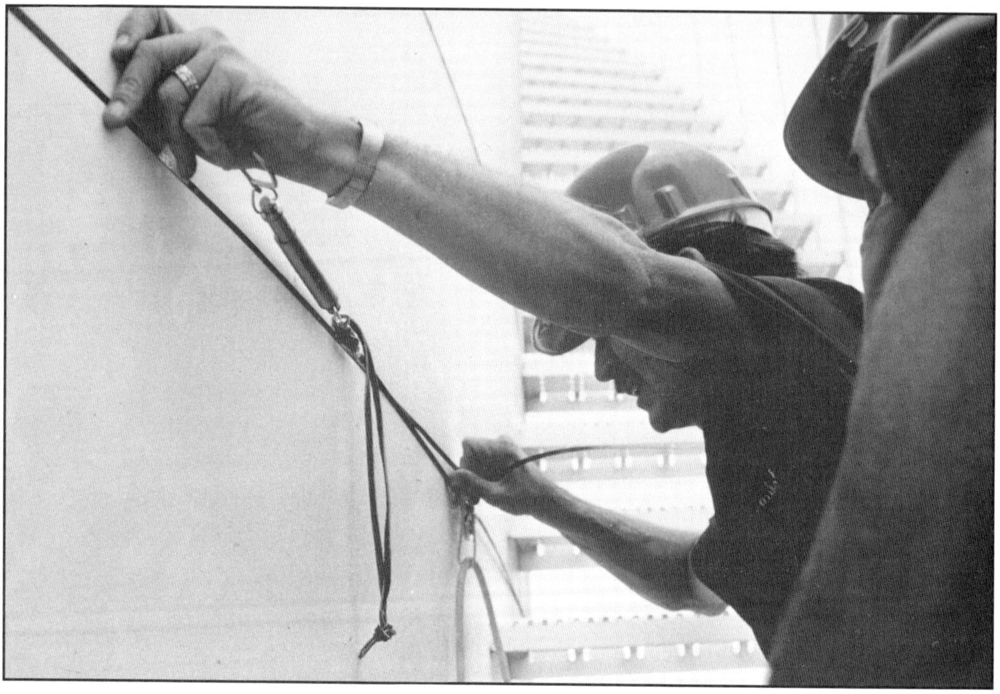

Figure 1.11. Applying tension to a working tape to calibrate it to the standard tape

Whether adding or subtracting a correction factor to each measurement, make sure the correction is not more than 1/100 foot per 100 feet of tape. If the correction is more than 1/100 foot per 100 feet, then the working tape is not within accepted tolerances and the crew must use a different one.

Another method of calibrating the working tape may be used if the working-tape circumference measurement is greater than the standard-tape circumference measurement. A crew member may apply tension on the tape until the working-tape length reads the same as the standard-tape length (fig. 1.11). In other words, if 184.625 is the measure by the standard tape and 184.635 is the measure by the working tape, then the crew member pulls on the working tape until it reads 189.625. A crew member records the amount of tension needed to get that reading. This tension reading may be different from the tension used on the standard tape, but it is the amount the crew must use to obtain all the remaining circumference measurements of that tank. Using this amount of tension calibrates the working tape to the standard tape.

MEASURING SUBSEQUENT RINGS

After calibrating the working tape, the crew is ready to measure the second ring. While standing on the stairs attached to the tank, crew members measure the width of the second ring using the now-calibrated working tape. They determine the tape's path and mark it on the tank with chalk or crayon and the strapping pole, just as they did on the bottom ring. (They use an extension on the pole to reach and mark the second ring.) After marking the tape path, they circle the second ring with the working tape (fig. 1.12). They check to be sure the tape is on the right path. Again they break the frictional resistance, or surface tension, between the tape and the tank shell, apply the required tension on the tape, and read it. They measure each successive ring in the same way until they reach the top ring.

The top ring is usually too high off the ground for the extension poles to reach, so a different technique is used. Instead of a strapping pole, the crew uses two ropes or lines and a metal ring to mark the tape path and to hold the tape as it circles the tank (fig. 1.13). First, while

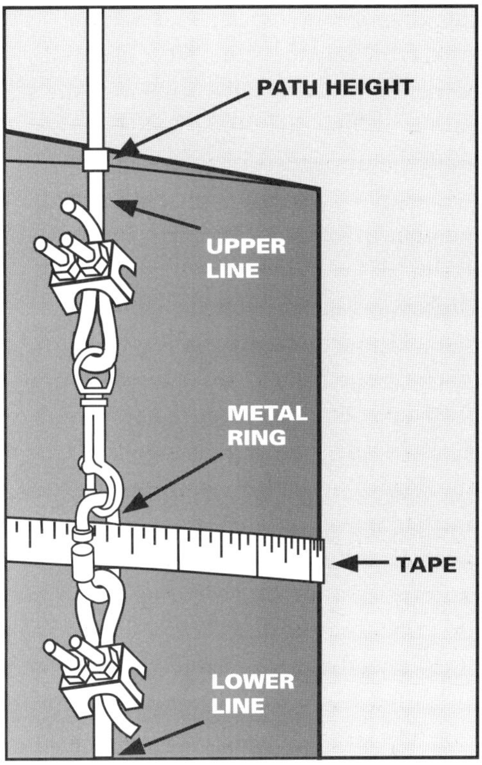

Figure 1.13. Tools for measuring top ring

Figure 1.12. Measuring the second ring

standing on the steps attached to the tank, crew members measure the width of the top ring and determine the tape's path from the top of the ring. Then they mark the upper line or rope to indicate the height of the tape path. One crew member then holds the upper line with this mark even with the top of the tank. The lower line is attached to the upper line with the metal ring. A crew member on the ground pulls on the lower line to keep tension on the upper line. A third crew member puts the tape through the metal ring, circles the top ring, and pulls the tape until it is taut. The person on top of the tank then checks the tape path at several points around the tank by sliding the upper line around the circumference of the tank, being careful to keep the tape path's mark aligned with the top of the tank. On larger tanks the crew can use a bo's'n's chair to make sure the tape is on the marked path. When crew members are certain that the tape is on the correct path, they remove the lines and ring, break the surface tension, and apply proper tension on the tape. They then record the measurement.

MEASURING SHELL HEIGHT

Once the circumferences of the tank's rings are measured and recorded, the crew measures the shell's height. The *shell height* is the distance between the bottom of the bottom angle of the tank and the top of the top angle (fig. 1.14). Crew members measure shell height near the hatch and at three other equidistant points around the shell. They record all four measurements and compare them. If these measurements differ by more than ½ inch, differences are accounted for when capacity tables are developed.

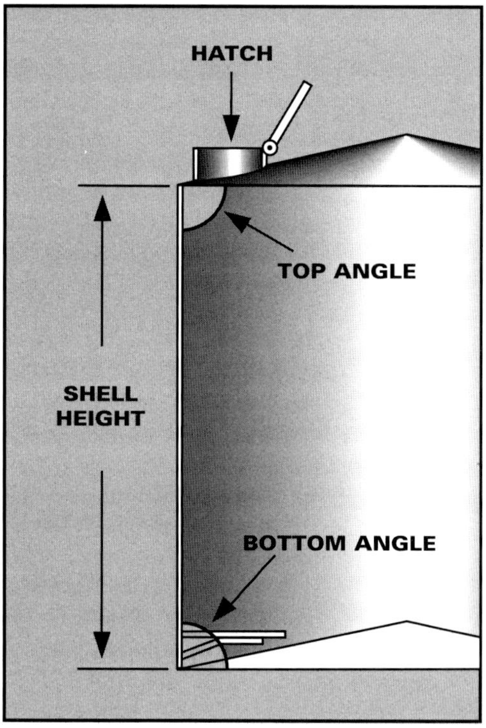

Figure 1.14. Measuring shell height

MEASURING GAUGING HEIGHT

Next the crew measures and records the *gauging height* (sometimes referred to as the *reference height*) of the tank, that is, the distance from a striking point on the tank floor or on a *datum plate* to a designated reference point on the gauge hatch.[2] To measure the gauging height, a crew member takes a gauger's tape (a metal tape marked in feet or metres with the smallest division in eighths of an inch or centimetres) and a bob (a graduated brass weight with a pointed tip) and lowers them into the tank until the tip of the bob touches the gauge striking point or the datum plate (fig. 1.15). When tape and bob are lowered into a tank, the bob keeps the tape taut so that the crew can get an accurate reading on the tape.

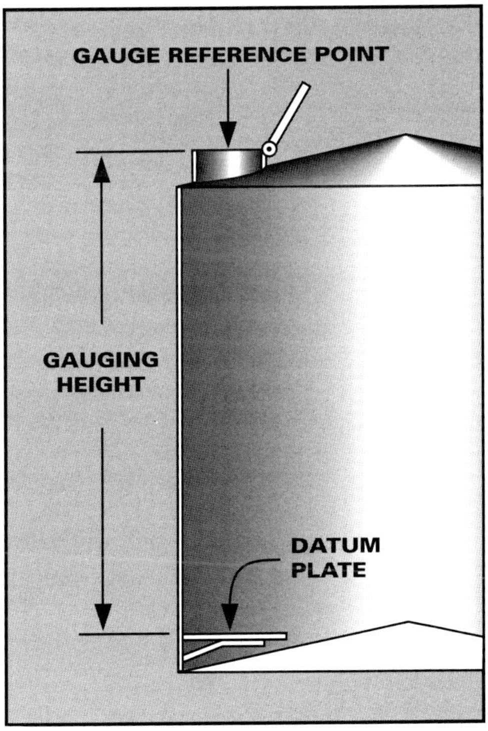

Figure 1.15. Measuring gauging height

[2] A datum plate is a level metal plate attached to the tank shell or bottom and located directly under the dipping reference point to provide a fixed contact surface from which liquid depth measurement can be made.

When the bob reaches the striking point or the datum plate, a crew member records the distance and, if any liquid is present, reads and records the height of the liquid in the tank by withdrawing the tape and noting where the liquid cuts the tape. The tank table department of the company must know the liquid height at the time of strapping in order to compute capacity tables accurately (why this is important is discussed later in this chapter, in "Measuring Temperature and Gravity").

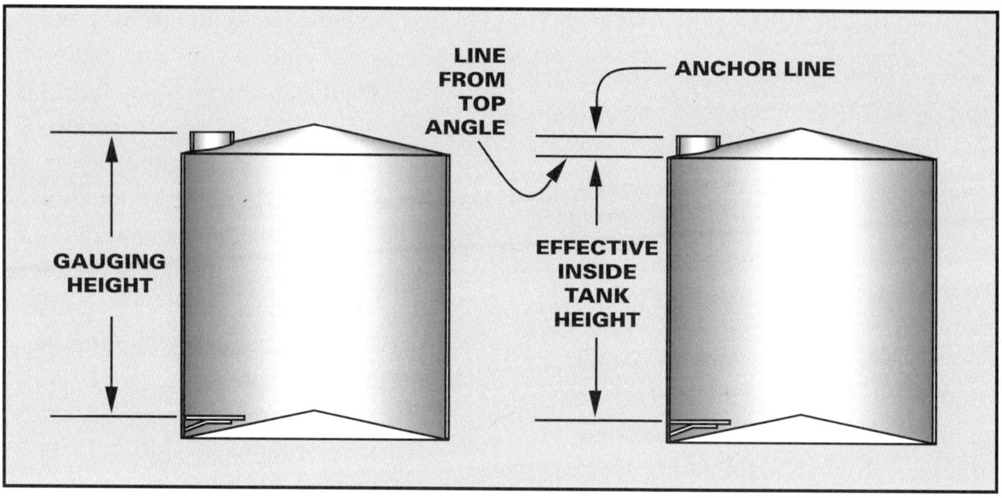

Figure 1.16. Measuring effective inside tank height

MEASURING EFFECTIVE INSIDE TANK HEIGHT

The crew also records the *effective inside tank height.* This height is the distance from the gauge striking point on the tank floor or the datum plate to the top of the top angle, or where the tank contents would begin to overflow (fig. 1.16). This height defines the upper and the lower limits of the capacity table and must be measured on the gauge path. Effective inside tank height is not the same as shell height, because the tank floor is not flat.

The crew begins by anchoring a line at the reference point on the hatch and extending it parallel to the top angle of the tank (fig. 1.17). A crew member levels the anchor line and then measures the distance between it and the top angle. The distance between the anchor line and the top angle is subtracted from the gauging height. The result is the effective inside tank height.

Figure 1.17. Anchoring line at reference point on hatch

MEASURING TEMPERATURE AND GRAVITY

As stated before, it is best to measure the capacity of a tank when the tank is empty and gas-free (i.e., free of hydrocarbon vapors). If a tank holds over 500 barrels and contains liquid at the time it is strapped, however, the crew is required to take a few more measurements: the height of the liquid in the tank, the temperature of the tank contents, the gravity of the contents, and the ambient temperature near the tank shell. These measurements are used in calculating a capacity table for tanks of 500 barrels or more. On these larger tanks the hydrostatic pressure of the liquid causes the shell to expand (fig. 1.18). This pressure is referred to as *liquid head stress*. As the liquid height in the tank changes, the liquid head stress changes. The gravity of the liquid and the temperature also affect the liquid head stress. This stress factor must be taken into account when capacity volumes are computed or the tables will not be accurate. The liquid head stress at the time the tank is strapped is thus computed first, then the tank's capacity is determined as if it were empty. Then the capacity is computed as if the tank were filled. The capacity table reflects shell expansion as each ring is filled. The capacity table will then be correct for any condition of fill. After crew members measure the temperature and the density of the liquid, they sign the report and the tank is strapped.

Figure 1.18. Liquid head stress causes tanks to expand.

FREQUENCY OF STRAPPING

How often a tank is strapped is determined by the parties involved. Any changes to the tank or any changes in the density of the liquids stored in the tank could alter its capacity. The type of service and the magnitude of the volumes stored in the tank can also determine how often the tank is strapped.

Some companies recommend that all tanks be strapped and new capacity tables computed every five years; API recommends every ten years. Even so, some tanks strapped in the 1920s have not been strapped since. If there is a dispute between the parties, the tank is strapped in the presence of witnesses from both sides.

CHAPTER 2

Gauging Petroleum and Petroleum Product Heights in Stationary Tanks

INTRODUCTION

Measuring oil and oil products stored in stationary tanks is a fairly straightforward process. First you strap, or measure, the capacity of the tank. Next, you develop capacity tables that tell you the gross volume, in barrels, for any level of liquid. Finally, you gauge (i.e., measure) the height of the liquid from the bottom of the tank to the surface of the liquid. You gauge the height of the liquid to take inventory or to measure the amount of liquid transferred any time liquid enters or leaves the tank. If liquid is moving into or out of the tank, you gauge the height before the liquid enters or leaves and after the liquid enters or leaves the tank. The difference between the two measurements is the gross volume transferred.

In order to determine as closely as possible how many barrels of liquid are in the tank, you gauge the height of the liquid to the nearest eighth of an inch. On crude oil lease tanks holding one thousand barrels or less, you gauge to the nearest quarter inch. Before you measure the liquid height, however, you need to make sure that the liquid has been in the tank long enough for the surface to be at rest and for any foam to have subsided. If you are gauging crude oil, you should also wait long enough for any entrained air to bubble out or water to settle out of the oil. And, if you are measuring liquid products and the tank has a mixer, you must be sure the mixer is shut off.

Once these conditions are met, you can use one of several methods to determine the liquid height in a stationary tank with a fixed roof: the *innage tape-and-bob procedure*; the *outage* (sometimes called *ullage*) *tape-and-bob procedure*; or a modified outage or ullage procedure. *Innage* refers to the liquid height in the tank. *Outage* refers to the space between the liquid and a reference point at the top of the tank.

INNAGE TAPE-AND-BOB PROCEDURE

The innage tape-and-bob procedure is a direct measure of the liquid in a tank. It is similar to measuring the oil in a car, where you take a graduated dipstick, wipe it clean, lower it into the crankcase, and withdraw it. You look at the end of the stick and note where the liquid has marked it. Depending on where the liquid *cuts* the stick, you know how much oil is in the crankcase. With a tank and the innage tape-and-bob procedure, you use a graduated metal tape and bob, make sure it is clean, lower it to the gauge striking point of the tank, withdraw it, and note where the liquid cuts the tape.

Since you're measuring a valuable liquid when you gauge a tank and even a slight mistake can cost the company a lot of money, you must be very careful and use the proper tools. For example, a typical metal tape and bob used in this procedure are of a particular shape and are graduated (some bobs are not graduated) (fig. 2.1). Most innage bobs are cylindrical with a pointed end. Starting from the zero point at the pointed end, the bob may be marked in ¼- or ⅛-inch graduations. The tape does not start again at zero, but continues where the bob leaves off. For example, if the bob measures 6 inches and the attached swivel is 2 inches long, the tape picks up with 9 inches. The bob and tape are marked this way so that you can easily read the height of the liquid from the bottom of the tank to the top.

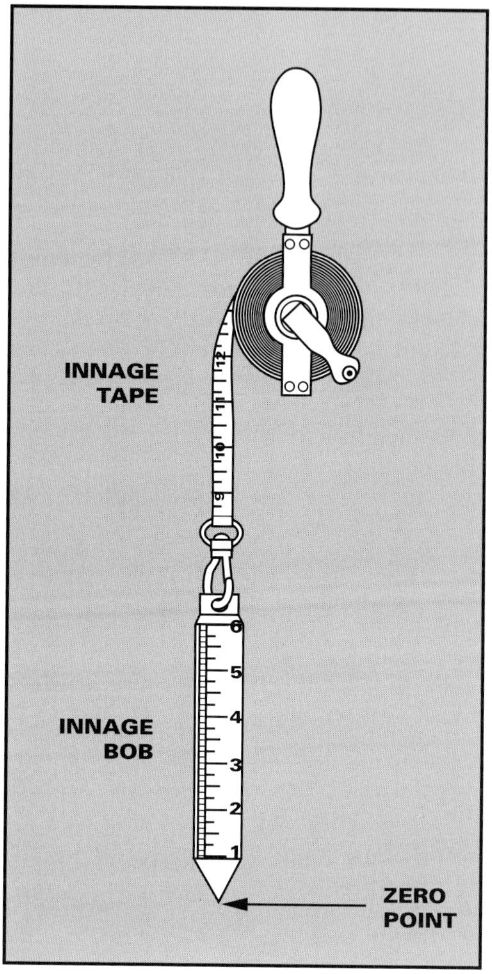

Figure 2.1. Innage tape and bob

The bob should be made of corrosion-resistant, nonsparking metal and be heavy enough to keep the tape taut. The bob tip should be sharp so that measurement will be accurate. The tape should be long enough to reach from the bottom to the top of the tank and not be kinked or spliced. All tapes and bobs should be checked for accuracy when new and at least annually thereafter. A working tape and bob should be calibrated against a master tape and bob that have been certified accurate at 60°F by NIST or calibrated against a certified NIST tape and bob. The zero point on the working tape and bob should check within 1⁄32 of an inch of the master tape with either both tapes in a vertical position and suspended with identical weights, or with both tapes pulled horizontally with a predetermined amount of tension.

When you are ready to gauge the liquid contents in the tank, you use two fixed points, or marks, on the tank. The first, the *reference point*, is usually in or on the hatch or gauging platform (fig. 2.2). All measuring starts from this point. All measurements end at the second mark, the *gauge*, or *zero point*, on the tank bottom, or on a *datum plate* located at the bottom of the tank directly beneath the reference point (fig. 2.2). The vertical distance between these two points is called the *gauging height*, or sometimes the reference height. This height is posted near the reference point and is a very important number. It is used as a guide in both the innage and the outage procedures. It also tells you how long the tape needs to be to reach the bottom of the tank. The gauging height must be verified against the gauging height listed on the capacity table computed for the tank (see chap. 1).

To obtain an innage gauge, you lower the tape and bob into the liquid until the bob is almost to the bottom (you use the gauging [or reference] height as a reference to know when to stop unreeling the tape). While lowering the tape, you keep the smooth, ungraduated side in contact with the rim of the hatch at the reference point (fig. 2.3). The metal-to-metal contact bleeds off any static electricity and prevents vapor ignition. Then you slowly unwind the tape until the bob just touches the bottom or the datum plate. The bob should touch only lightly so that it will not tilt or deviate from vertical. An experienced gauger can tell by the feel of the tape when the bob touches bottom.

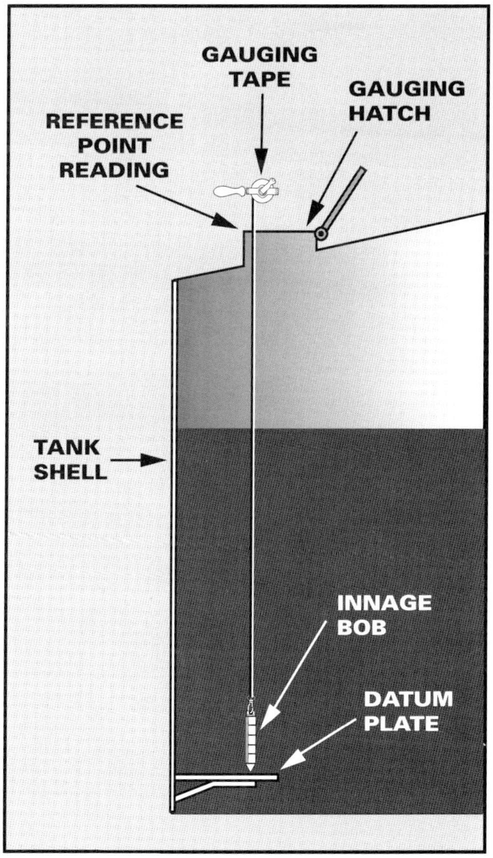

Figure 2.2. Fixed measuring points for liquid contents

Figure 2.3. Obtaining the innage gauge

Compare the tape reading with the posted gauge height. If the tape reads less than the gauge height, there may be a buildup of sediments on the tank bottom and another method of measurement (usually the outage tape-and-bob procedure) may be necessary. If the tape reads more than the gauge height, the tape may not be vertical. In either case, withdraw the tape, clean it, and lower it into the tank again. If the tape reads the same as the posted gauge height, pull up the tape and read the height where the liquid cuts the tape (fig. 2.4). This is the liquid height in the tank. Gauge again to verify the reading. If both readings are the same, immediately record the result. You need two identical readings to verify the results.

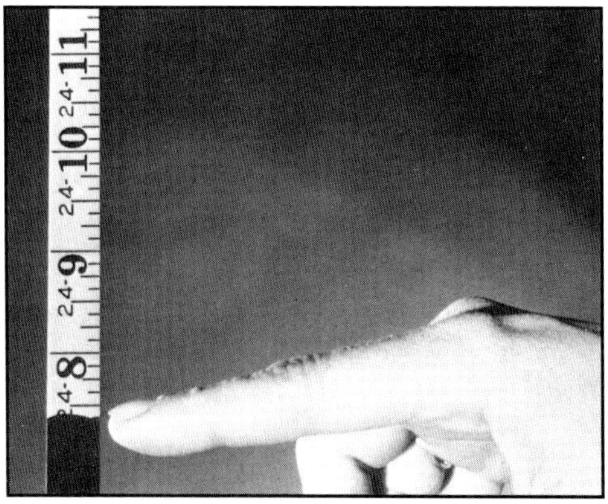

Figure 2.4. Liquid cut on tape

OUTAGE TAPE-AND-BOB PROCEDURE

Sometimes it is impossible to measure the liquid height in a tank directly. There may be so much sediment buildup—that is, tank bottoms—that the bob cannot touch the bottom of the tank; or the tank bottom may flex, causing varying gauge heights; or the liquid may be very thick (like asphalt), so getting a vertical line to the bottom is impossible. Whatever the reason, you can use an indirect method of measurement called the outage (or ullage) tape-and-bob procedure. With this procedure, you do not directly measure the liquid level; instead, you measure the space between the top of the liquid and the reference point on the tank (fig. 2.5).

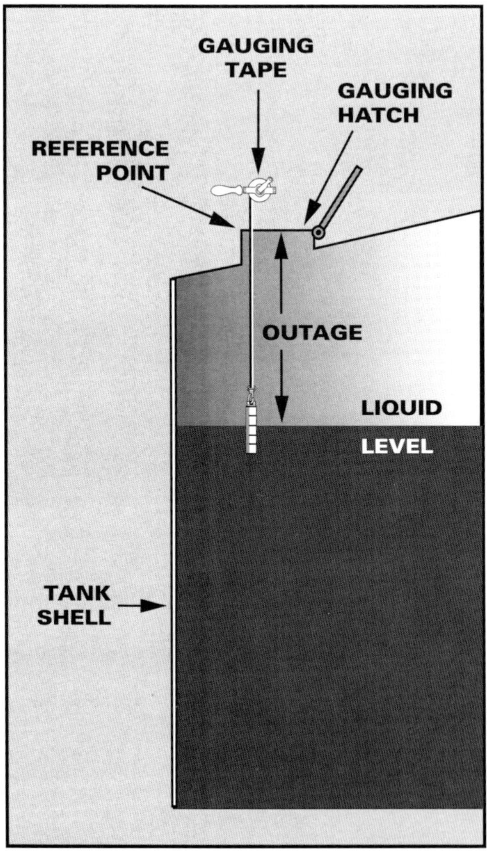

Figure 2.5. Indirect measurement of liquid level

You then determine the liquid height by subtracting the outage measurement from the gauging height. For example, if the gauging height of the tank is 46 feet, 6¼ inches, and you determine that the outage is 24 feet, 3 inches, then the liquid height in this particular tank is 22 feet, 3¼ inches (fig. 2.6).

The tape and bob used in this procedure is different from the one used with the innage procedure. As you can see in figure 2.7, the zero point on the outage bob falls at the swivel between the tape and the bob. The outage bob is also usually three- or four-sided instead of cylindrical and has a flat tip.

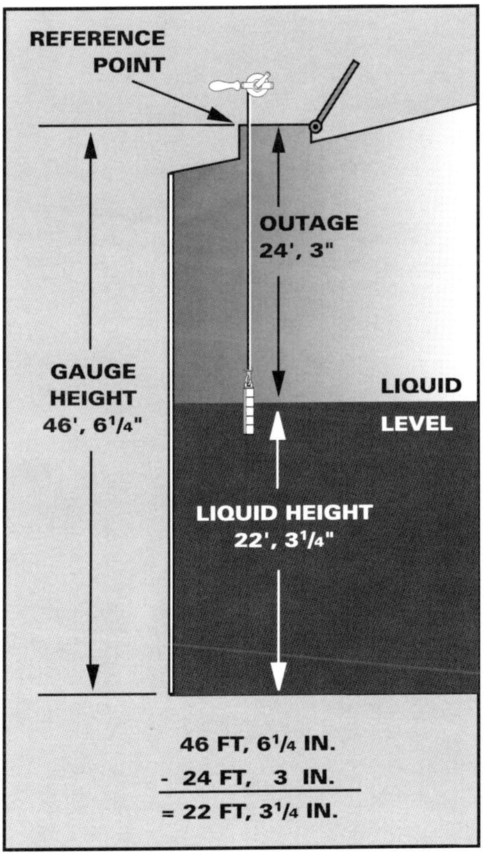

Figure 2.6. To determine liquid level, subtract the outage from the gauge height.

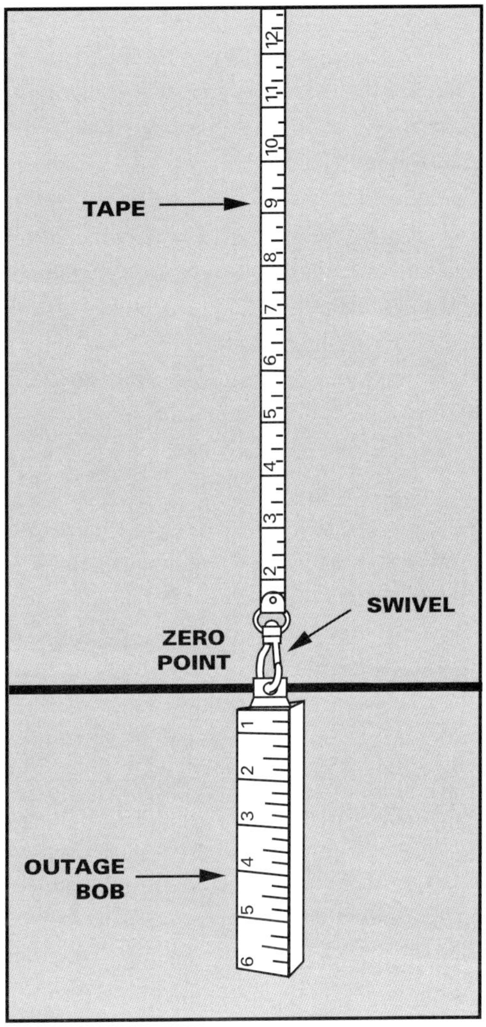

Figure 2.7. Zero point on an outage bob

In spite of these differences, there are some similarities between the two instruments. Like the innage tape, the outage tape should be of corrosion-resistant metal with a heavy enough bob to keep the tape taut. The outage tape also needs to be long enough to reach the bottom of the tank and not be kinked or spliced. It too should be checked for accuracy and calibrated with a master tape before the first use and recalibrated annually thereafter.

To obtain an outage measurement, you start from the reference point on the tank and slowly lower the tape and bob until the bob just touches the surface of the liquid. (Again be sure the smooth, ungraduated side of the tape stays in contact with the side of the hatch.) After the bob has stopped swinging, lower the tape slowly until several inches of the tape or the bob are submerged. This allows you to get a definite liquid cut on the tape or the bob. Record the reading of the tape at the reference point. Withdraw the tape and record the liquid-level cut on the tape or bob. Add the tape reading at the reference point to the bob reading to get a measure of the outage for that tank. For example, suppose the tape reading at the reference point is 19 feet, 4 inches and the reading on the bob is 2¼ inches. Adding these two figures together you find that the outage measurement is 19 feet, 6¼ inches. To determine the liquid height, subtract the outage measurement from the gauging height. Assuming a gauging height of 46 feet, 6¼ inches, the liquid height in the above example is 27 feet (46 feet, 6¼ inches minus 19 feet, 6¼ inches) (fig. 2.8).

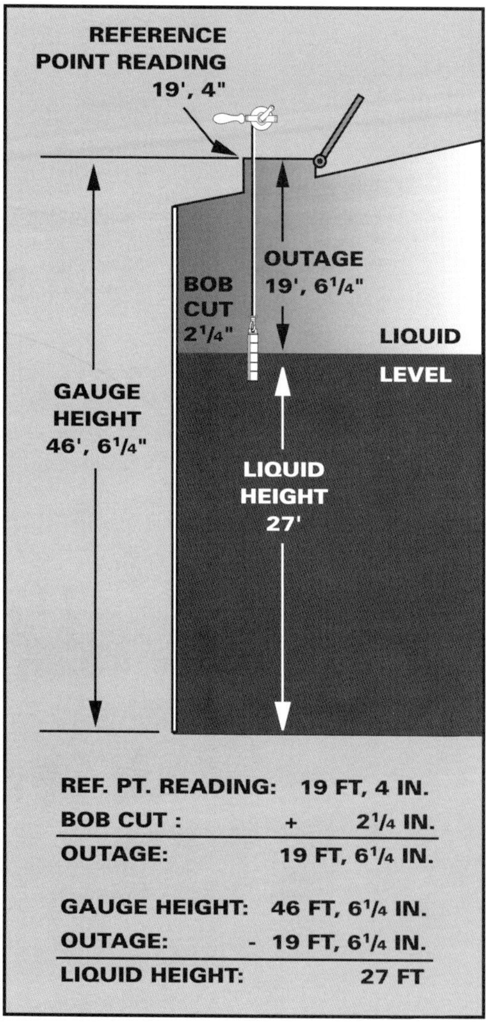

Figure 2.8. Outage tape-and-bob procedure

MODIFIED OUTAGE PROCEDURE

With this procedure, you do not find an outage sum; instead, you use an outage tape and bob to find an innage measurement. You lower the outage tape and bob until the bob is completely submerged, record the tape reading at the reference point, withdraw the tape and bob, read the liquid-level cut on the tape, and record this reading. Subtract the tape reading at the reference point from the gauging height. Add the difference to the liquid level reading on the tape. The result is the liquid-level height (innage measurement). For example, suppose the tape reading at the reference point is 32 feet, $5\frac{7}{8}$ inches, and the liquid-level reading on the tape (the tape cut) is 1 foot, $2\frac{3}{8}$ inches. Assuming the gauging height is 43 feet, $5\frac{7}{8}$ inches, the height of the liquid in the tank is 12 feet, $2\frac{3}{8}$ inches (fig. 2.9).

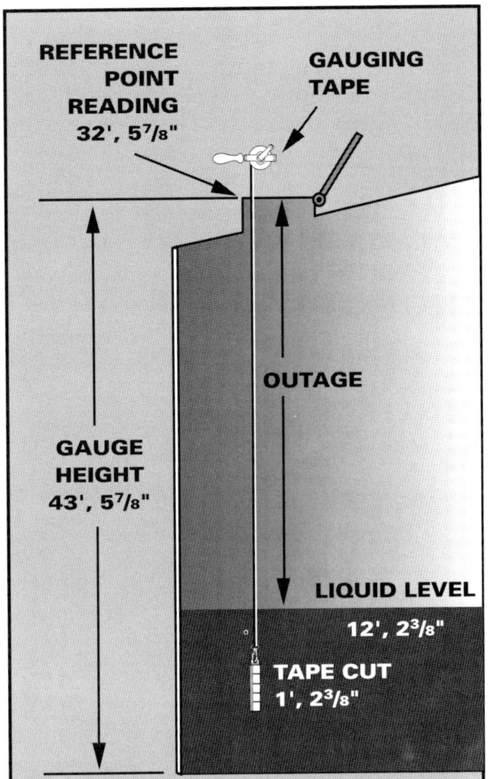

GAUGE HEIGHT:	43 FT, $5\frac{7}{8}$ IN.
REF. PT. READING:	- 32 FT, $5\frac{7}{8}$ IN.
DIFFERENCE:	11 FT
TAPE CUT:	+ 1 FT, $2\frac{3}{8}$ IN.
LIQUID LEVEL:	12 FT, $2\frac{3}{8}$ IN.

Figure 2.9. Modified outage procedure

COMMON PRACTICES

Regardless of which method of measurement you use, there are certain practices you should follow whenever you gauge a tank.

Equipment Inspection

To start, you should always inspect the equipment before gauging. Make sure the point on the innage bob is sharp. A dull bob tip will lead to innaccurate innage gauge readings. All gauge tapes should be clean, dry, and in good condition. Wet tapes will not give a good liquid level reading, rusty tapes are difficult to read, and kinked or spliced tapes give inaccurate results.

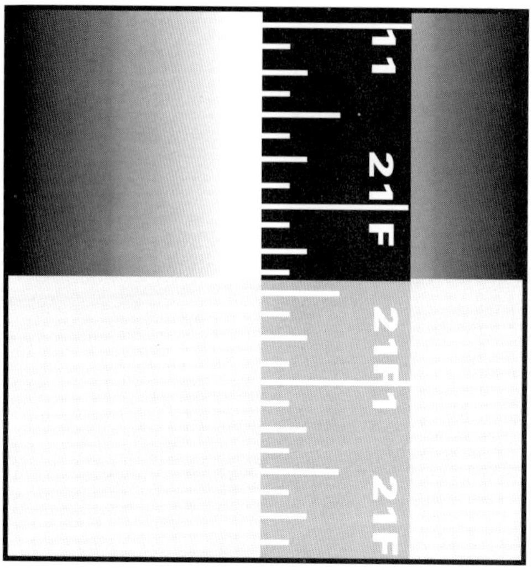

Figure 2.10. Black tape

Colored Tapes

You should use different-colored tapes for different types of liquids. Dark liquids, such as crude or heavy fuel oils, show up best on chrome-clad or blond tape. Dark liquids evaporate slowly, giving you plenty of time to observe the cut (fig. 2.4). Clear liquids, such as kerosene, turbine fuels, light fuels, solvents, and chemicals, show up best on black tape (fig. 2.10). Even so, some clear liquids, like gasoline, condensates, and other light-end products, evaporate so quickly that the cut disappears before you can read the tape. With these products, you should use a product-finding paste on the tape (fig. 2.11). Never use chalk for this purpose; the liquid in the tank has a tendency to creep up on chalked tape and give an inaccurate reading.

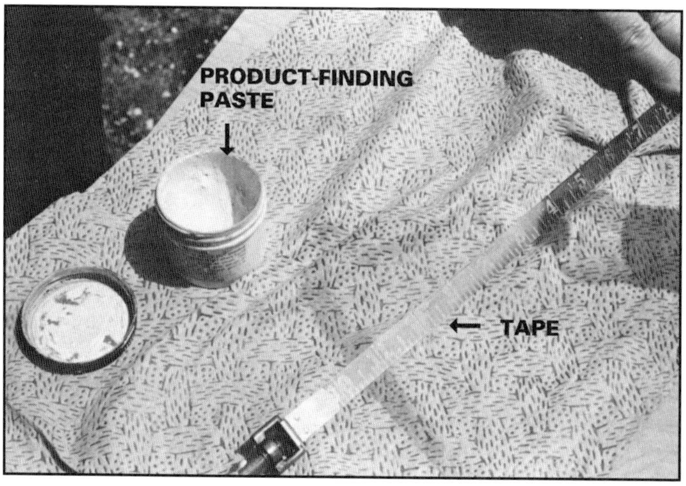

Figure 2.11. Product-finding paste on a measuring tape

Gauging before Custody Transfer

If you are gauging before a custody transfer, you isolate the tank and all lines. To do this, fill all lines from the tank to the point of transfer and close the valve at that point. This stops all movement into and out of the tank during the gauging operation. If you are taking tank samples in the gauge path, be sure to do this before you gauge. Allow adequate time for any sediment and water (S&W) to settle and for the surface of the liquid to calm before gauging. Finally, always gauge the tank until you have two identical measurements or three measurements within an eighth of an inch of each other. Even though you perform several measurements to ensure accuracy, the results are referred to as a single gauge. For example, during a custody transfer, the measurements done before any liquid enters or leaves the tank are called the *opening gauge*, the measurements performed after liquid enters or leaves the tank are called the *closing gauge*.

Gauging for Inventory

If you are gauging for inventory purposes and it is impossible to isolate the tank, be sure to gauge at the same time of day as the last inventory gauge. Again, if you are taking tank samples, be sure to take them before you gauge.

If the tank has a ground-level or remote-reading gauge, record the reading on this gauge so it can be compared with the measurements you get.

Safety Measures

Once you are ready to begin gauging the tank and you start up the stairway, be sure you ground yourself by keeping one hand on the metal handrail. When you open the hatch, stand upwind and keep your face turned away to avoid breathing the vapors from the tank. (If you are gauging in a hydrogen sulfide area, be sure to wear the proper breathing equipment and to take the necessary precautions before gauging.) Wait until any pressure buildup has dissipated before you begin to gauge.

As you lower the tape and bob into the tank, remember to keep the tape in contact with the tank metal. This contact will bleed off any static electricity and keep the vapors in the tank from igniting. If the temperatures of the bob and the liquid differ considerably, the surface of the liquid may boil when the bob comes in contact with the surface. Always wait until the boiling stops before lowering the bob further.

GAUGE PIPES

Some tanks are equipped with gauge pipes through which all gauging must take place. Some of these pipes are slotted throughout their full length (fig. 2.12), while others are slotted only at the bottom. A pipe that is slotted its full length allows free transfer of liquid between the tank and the pipe. Because of this free transfer, the liquid in the pipe is the same temperature and density as the liquid in the tank, so gauging through the pipe gives you an accurate measure of the liquid height. (Remember, temperature and density affect the volume of oil and oil products.)

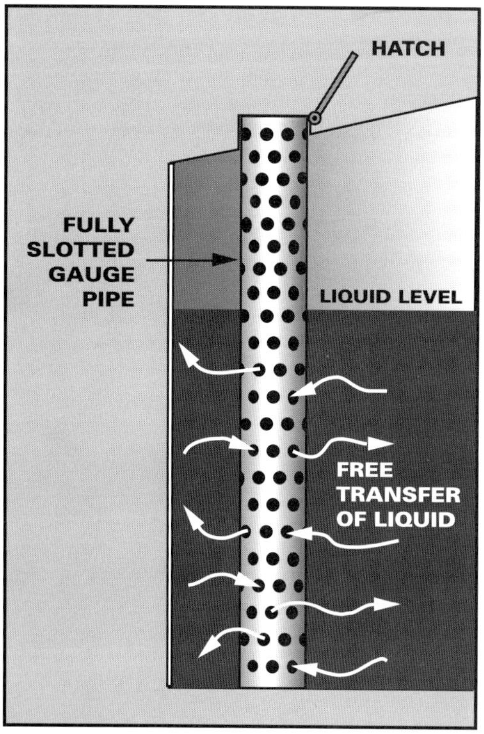

Figure 2.12. Gauge pipe slotted throughout its length

Some pipes, however, are slotted only at the bottom. With a pipe slotted only at the bottom the slots must be high enough above the crude oil / free water interface to permit the free transfer of liquid. If the slots are too low there may not be a free transfer of liquid and the liquid in the pipe may not be the same as the liquid in the rest of the tank (i.e., the liquid that moves in and out of the tank may not always be of the same gravity or temperature as the liquid already in the tank). The result is inaccurate measurement. Whenever you gauge through a pipe, use a mirror or approved light to see if the pipe is slotted above the liquid surface. If the slots are not visible, ask your supervisor for further instructions.

FLOATING-ROOF TANKS

Some tanks have roofs that float on the surface of the liquid (fig. 2.13). Floating roofs form a seal on top of the liquid and help minimize vapor losses, so they are used when storing volatile liquids such as gasolines, naphthas, and some crude oils. To gauge the liquid heights in these tanks, you use the same procedures described for tanks with fixed roofs.

If federal, state, and company safety regulations all permit, you can gauge a floating-roof tank from the deck of the roof. If you do so, use only the innage method, since you do not have the advantage of the gauging height as a reference. If you measure from a gauging platform, you can use either the innage or the outage method. No matter which method you use, however, there are several factors you must take into consideration before you begin.

First, the roof must be floating freely and not be hung up on any part of the tank. If it is not floating freely, it could distribute the liquid unevenly

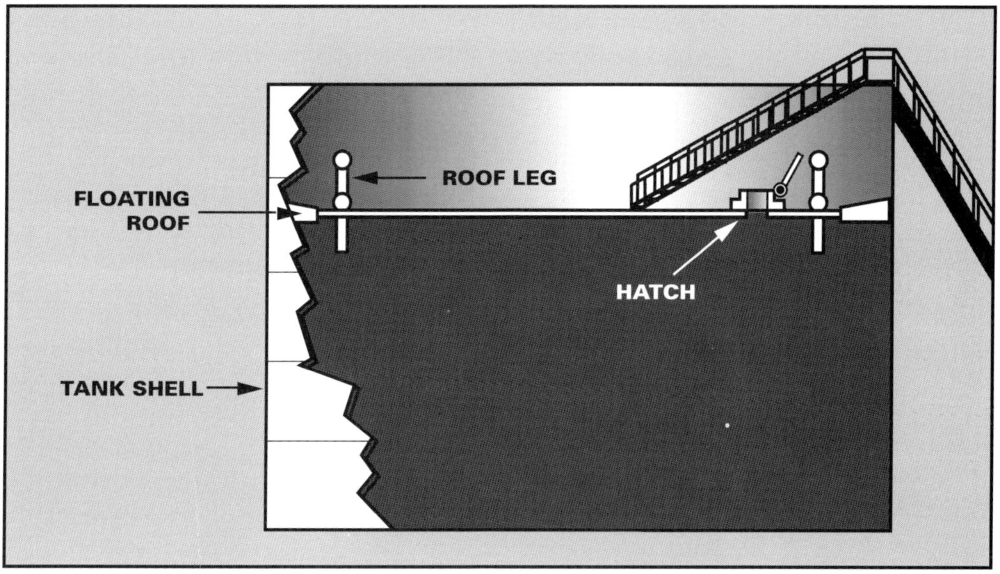

Figure 2.13. Floating-roof tank

throughout the tank and cause inaccurate measurement of the liquid height (fig. 2.14). If, however, the liquid level has dropped so low that the roof is completely supported by its legs (i.e., the legs are resting on the bottom of the tank), you can measure the liquid height accurately. With the legs resting on the tank bottom, the liquid should be evenly distributed. Just make sure that the legs are resting on the tank bottom and not on any buildup of S&W. The accumulation of S&W is usually uneven, so if the roof is sitting on anything but the bottom of the tank, it will not be level. If an S&W buildup is evident, record this information and include it in your gauging report.

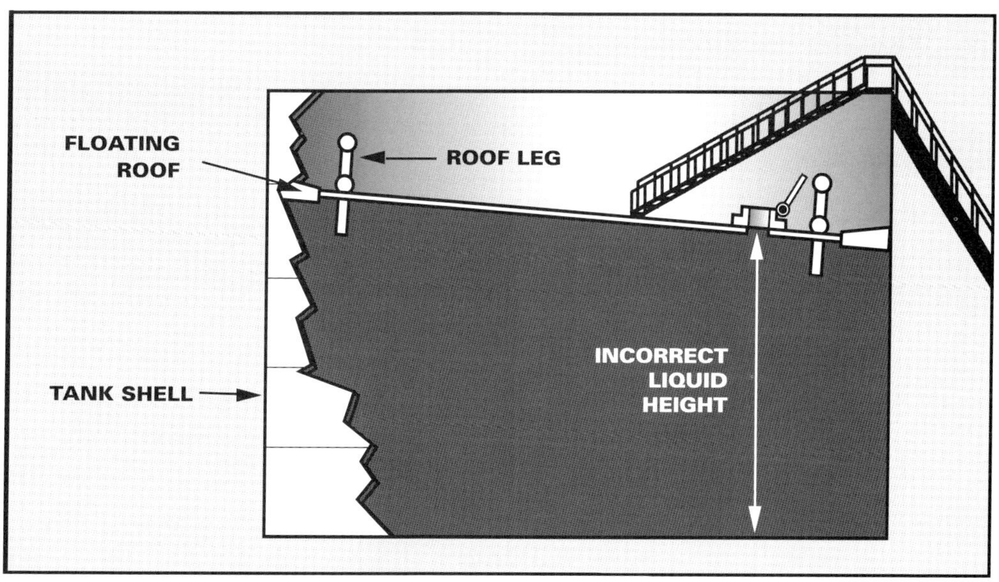

Figure 2.14. Unevenly floating roof

With most floaters, the legs can be in one of two positions: in the high position, the roof is approximately 6 feet above the tank floor when the legs are resting on the tank bottom; in the low position, the roof sits approximately 3 feet above the tank floor when the legs are resting on the tank bottom. Gauge tables are usually developed from measurements taken when the legs were in the low setting. The roof starts to float when the liquid reaches the underside of the roof (3 feet). The entire roof does not lift off the floor all at once; it takes approximately 6 additional inches of liquid added to the tank before the entire roof floats freely. If possible, tank gauging should not take place if the liquid height is between 3 feet and 3 feet, 6 inches (i.e., in the *critical zone*), because all of the roof may not be afloat and you will not be able to measure the liquid height accurately. The critical zone is noted in the tank's gauge table, but the volumes listed in the critical zone are not necessarily accurate. The only way to obtain accurate volumetric data for a tank table in this zone is to use the liquid calibration procedure, described in API Standard 2555, which is part of the *MPMS,* chapter 2.

The weight of the floating roof is another important factor. Because the roof rests on top of the liquid, it displaces a certain amount of liquid around its edges and up into the tank hatch (fig. 2.15). The amount of liquid displaced varies, depending on the gravity of the liquid in the tank and the weight of the roof. When you gauge the liquid height, you must take this displacement into consideration. Gauge tables for floating-roof tanks adjust for liquid displacement. In table 2.1 you can see that, for this particular tank, roof #270,400 displaces 914.76 barrels of liquid when completely floating on 35.9 API degree oil.

During gauging, make sure the roof weight remains constant, because changes in roof weight alter the displacement values. For instance, accumulated rainwater increases roof weight and causes greater displacement. If there is any water on the roof, drain it (by using available siphon equipment or by calling in a vacuum truck) before gauging.

Scale and dirt also increase roof weight, but the greatest weight change is caused by accumulated snow and ice. If the floating roof has snow, ice, or any other debris on it that cannot be drained easily or cleaned and you are gauging before and after a custody transfer, make sure that both gauges take place before the roof drops so low that the legs touch bottom. Remember that the weight of the roof must be the same before and after the liquid in

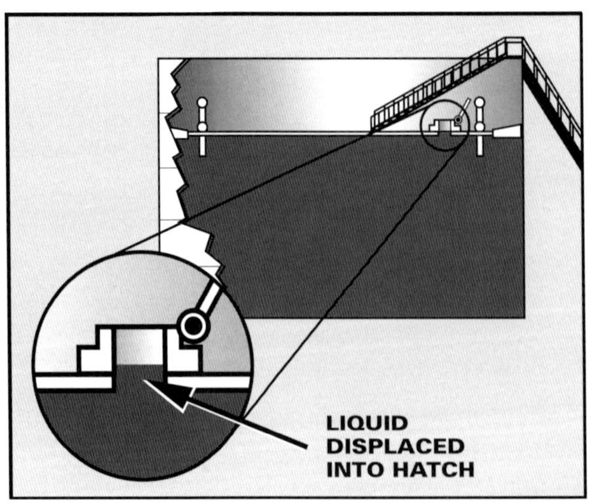

Figure 2.15. Weight of floating roof displaces some liquid into hatch.

TABLE 2.1
Excerpt from a Capacity Table for Floating-Roof Tanks, Roof on Low Legs

4		9		14		19		24		
1	8788.61	1	19705.33	1	30626.13	1	41553.24	1	52480.18	1
2	8970.50	2	19887.35	2	30808.14	2	41735.37	2	52662.29	2
3	9152.40	3	20069.36	3	30990.16	3	41917.49	3	52844.40	3
4	9334.29	4	20251.37	4	31172.28	4	42099.61	4	53026.50	4
5	9516.18	5	20433.39	5	31354.40	5	42281.73	5	53208.61	5
6	9698.08	6	20615.40	6	31536.52	6	42463.86	6	53390.72	6
7	9879.97	7	20797.41	7	31718.65	7	42645.98	7	53572.83	7
8	10061.86	8	20979.43	8	31900.77	8	42828.10	8	53754.94	8
9	10243.76	9	21161.44	9	32082.89	9	43010.22	9	53937.05	9
10	10425.65	10	21343.45	10	32265.01	10	43192.34	10	54119.16	10
11	10607.54	11	21525.47	11	32447.13	11	43374.47	11	54301.27	11

Liquid head correction per API Standard 2501, with supplements. 270,400-pound roof displaces 914.76 barrels in 35.9 API degree oil. Avoid *marked* area; roof not metered.

the tank is delivered. If the roof is supported by the legs, it no longer displaces liquid around its edge or up into the hatch. Also, if the ice and snow begin to melt, do not drain the water until the custody transfer is complete.

The wind can also affect the liquid level in tanks with floating roofs. It can force the roof up and down and cause the liquid level to change, thereby making it impossible to get repeated liquid cuts in the same place on the tape. So on windy days, gauge several times and average the results.

Some tanks with fixed roofs have lightweight internal covers that float on the liquid (fig. 2.16). These covers are more flexible than conventional floating roofs. Because they are light, these covers tend to flex, with parts lagging during high fill or withdrawal rates. Always allow plenty of time for them to stabilize before gauging.

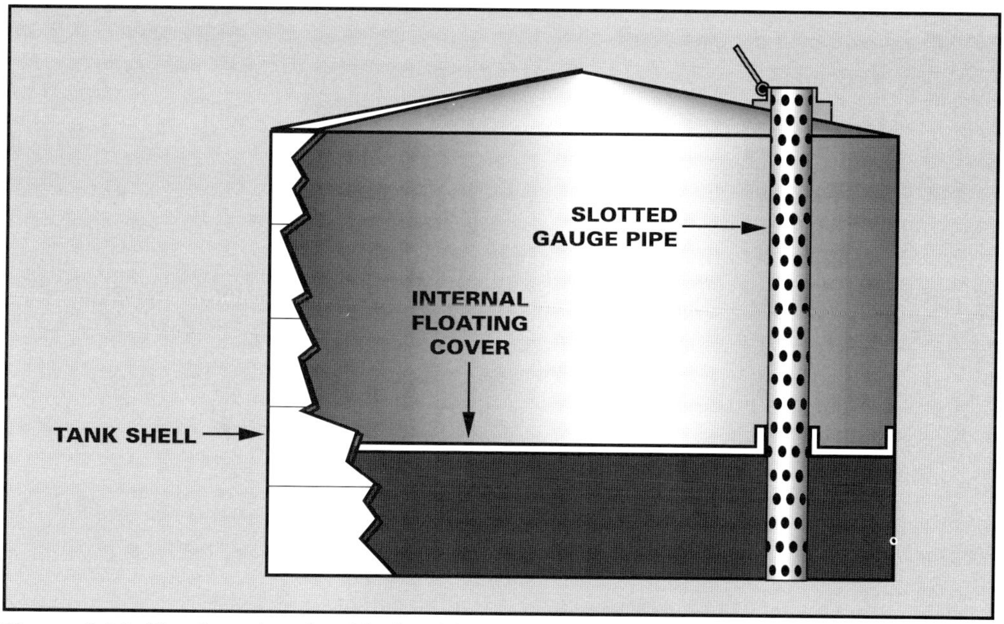

Figure 2.16. Fixed-roof tank with flexible internal floating cover

GAUGING PRESSURE TANKS

Volatile liquids such as liquefied petroleum gas and finished products such as natural gasoline, blending stocks, and aviation and motor gasolines are stored in pressurized tanks that cannot be opened for gauging. With these tanks you must use the installed internal gauging equipment (fig. 2.17). You can gauge pressure tanks with the same innage, outage, or modified outage procedures described earlier. The techniques, however, differ somewhat because the equipment is fully enclosed. All internal gauging is performed mechanically through a series of locks and valves, and the liquid cut is read through a glass window.

Some internal tape-and-reel assemblies are equipped with a tension lock that automatically locks when the bob touches bottom. It is more difficult to use those without the lock mechanism because you cannot feel when the internal tape and bob touch bottom. You must use the gauging height posted at the reference point to estimate how much tape to unwind. When the tape measure indicates that you should be at the bottom or at the datum plate, you check the tape through the window to make sure it has no slack in it. If you do see slack, reel the tape in and try again. If repeated tries indicate slack, you may need to use the outage or modified outage procedure to gauge the tank. If there is no slack in the tape when you unwind the tape to gauging height, reel in the bob and read the liquid cut on the tape. If you are gauging with a system that automatically locks when it touches bottom and you notice slack in the tape, report the tape and bob for repair.

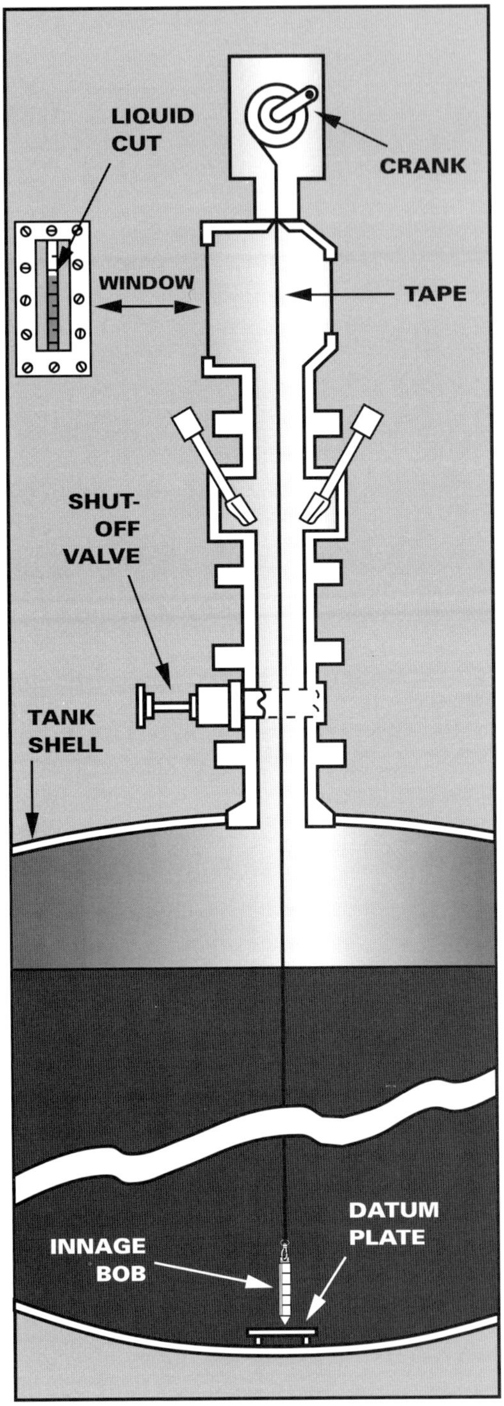

Figure 2.17. Internal gauging equipment

In addition to reporting any problems you find with the tape and bob, you should also report any vapor you see or smell around the valves or chamber door.

If you are gauging with the outage or modified outage procedure, the tank must have a sight glass on the side or a floating gauge so that you can tell when the bob touches the liquid. If neither of these is available, you need to have an estimate of the liquid height so you can tell when to stop unreeling the tape in order to get a clean cut on the bob.

Some pressure tanks are equipped with either an automatic gauge or a gauge glass. Instead of physically gauging tanks with these mechanisms, you read and report the liquid height indicated by the gauge. You should calibrate automatic gauges in accordance with the API *MPMS*, chapter 3, section 1, by performing several hand-line gauges or by gauging through available internal tape-and-bob measuring equipment.

If you are gauging the tank with a direct reading gauge glass (fig. 2.18) installed on the tank, you first draw off some liquid to make sure that the temperature of the liquid in the glass is now the same as the temperature of the liquid in the tank. Then you use a spirit level or a straight edge (fig. 2.19) to read the liquid level in the glass and to obtain an accurate reading of the liquid height in the tank.

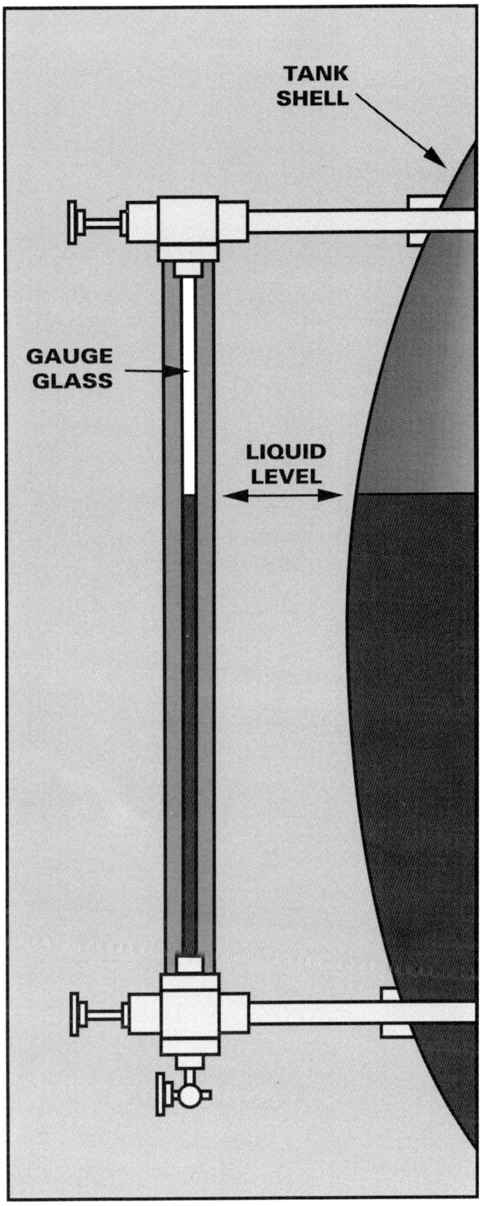

Figure 2.18. Pressure-tank gauge glass

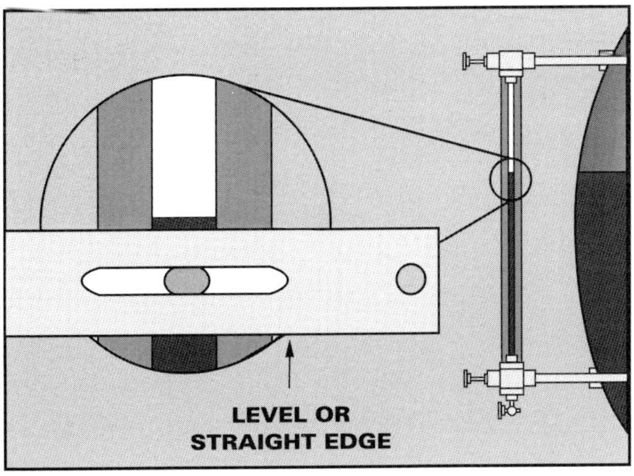

Figure 2.19. Straight edge used to gauge pressure tank

GAUGING SETTLED S&W

When you gauge liquid height, you are determining the gross amount of oil or product in a tank. In order to obtain an accurate measure, that is, the net amount of oil or product in the tank, a few other tests must be performed. The amount of settled sediment and water in the bottom of the tank must be measured. The amount of settled S&W is then subtracted from the total amount of liquid in the tank. After other tests (such as those for determining suspended S&W, specific gravity, and temperature) are completed, the net volume of oil or product can be determined.

The most common method of measuring the height of settled S&W in the tank is the water cut. The water cut is usually performed at the same time as the innage gauge. (If you are using one of the other procedures to determine liquid height in the tank, then determining S&W height is a separate procedure.) To obtain a water cut, before lowering the tape and bob into the liquid you apply a thin, smooth layer of water-finding paste to the area of the tape where you expect to find the water cut. Water-finding paste is a substance that reacts with the settled water and either dissolves or turns red, depending on the type of paste. Be careful to apply the right amount of paste (determine the amount by trial and error): too much will run and distort the reading; too little will not show a clear, readable water cut.

As you lower the tape and bob into the liquid, the water-finding paste will become covered with oil or product. When the bob touches bottom, allow enough time for the water to penetrate the oil or product and react with the paste. In general, the heavier the liquid, the longer the immersion time. For example, if you are gauging a tank of gasoline, you leave the bob at the bottom for about 10 seconds before reeling it back to the surface. If you are gauging a tank with crude oil, wait approximately 1 minute. Just be sure the water has time to react with the paste.

Note the height of the reference point and reel in the tape. Subtract the water-cut measurement from the overall height of the liquid column. For instance, suppose you determine that the liquid height in the tank is 36 feet, 4 inches. Lower an innage tape and bob to the gauging point, reel it back in, and note the water cut, in this case, 2¼ inches. Subtract 2¼ inches from 36 feet, 4 inches to determine that the height of the liquid column less S&W is 36 feet, 1¾ inches.

If you are gauging in rain or heavy fog and the water-finding paste washes off or changes color before you can lower the tape, you should wait until the rain or fog clears before gauging. If you are gauging the S&W level in crude oil, you may need to follow a few special procedures. These procedures may deviate from standard gauging practices and so must be approved by all parties involved. First, apply a layer of light lubricating oil over the water-finding paste. Next, lower the tape and bob to the bottom and allow a full minute for the water to penetrate the oil and react with the paste. When you reel in the bob, you may find crude oil obscuring the water cut. If so, rinse the tape or bob with solvent to remove the oil. The solvent removes the oil but leaves the paste.

Sometimes the buildup on crude oil tank bottoms is particularly heavy and the innage bob may not be able to penetrate it. In this case, you can gauge the S&W with a heavier and longer water gauge bob (fig. 2.20). Since the water gauge bob is longer than the ordinary bob, you must mathematically adjust the tape reading at the reference point and at the liquid cut line to account for the difference.

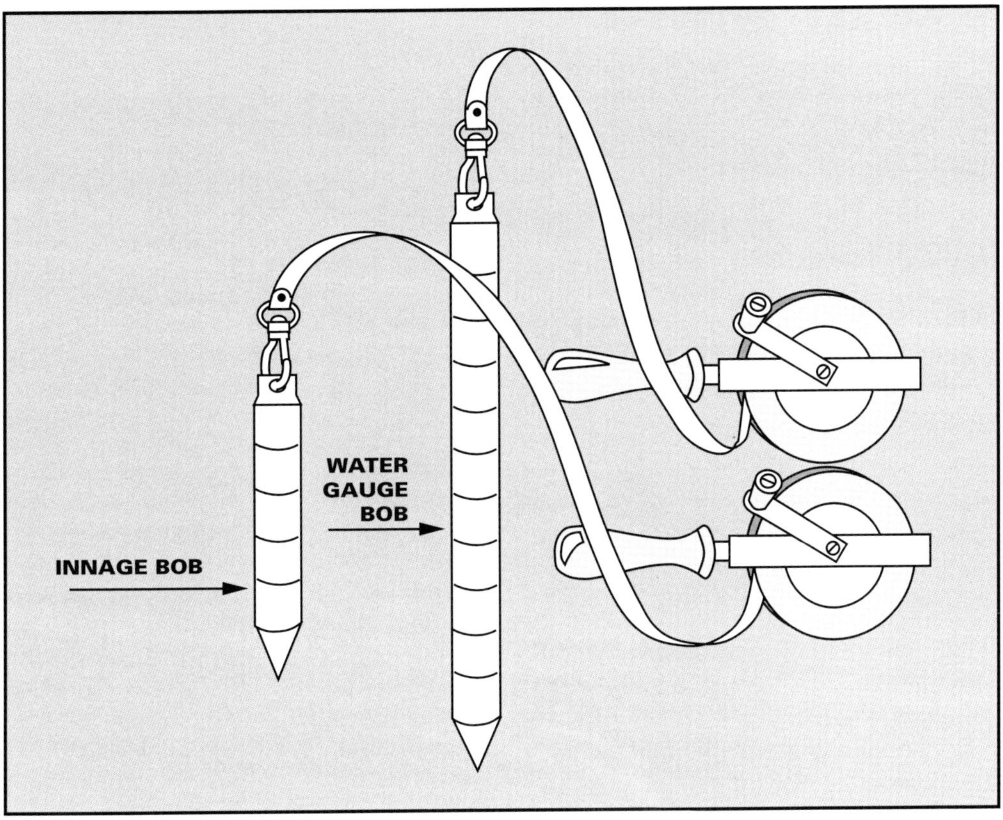

Figure 2.20. Water gauge bob

Another way of determining the S&W height is to use a core thief (fig. 2.21). After you have checked the thief to be sure that it is clean and in good working condition, open the bottom closure, attach the thief to a rope, and lower it into the tank at the reference point. Be sure that the thief comes into contact with the bottom slowly, so that the closure valve is not tripped. Raise the thief slightly and lower it more quickly, so that it will shut when it hits the bottom of the tank. Raise the thief to the surface and pour the contents back into the tank until water globules appear in the thief liquid or the liquid becomes discolored. Set the thief in an upright position and measure the height of S&W. This is the approximate height of S&W in the tank. If the trip rod on the thief was extended, the amount of extension must be added to the S&W height in the thief.

Figure 2.21. A core thief

For example, if the S&W reaches the 5-inch mark in the thief and the trip rod extends 6 inches beneath the thief, then the height of the S&W is 11 inches, not 5 (fig. 2.22).

CONCLUSION

Since determining the liquid height is so important (small errors can mean thousands of dollars), you should make sure that you gauge each tank carefully with tools in good condition and following the procedures described in this chapter and the API *MPMS*, chapter 3, section 1. You should also gauge more than once to verify your results and then write the results down. Do not rely on memory. Once you determine the gross amount of liquid, you will be ready to perform the tests necessary to measure accurately the net amount of oil or product in a tank.

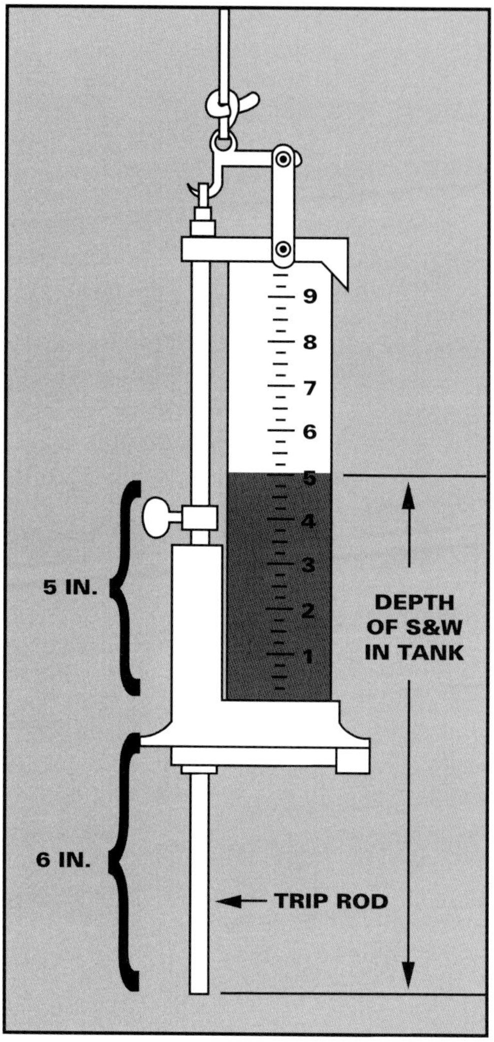

Figure 2.22. Core thief with trip rod extended

CHAPTER 3

Measuring the Temperature, Density, and Suspended S&W Content of Liquids in Tanks

INTRODUCTION

To determine the net volume of liquid in a tank, you must know not only the height of the liquid and the settled S&W, but also the liquid's temperature, its density, and the amount of S&W suspended in it. All three of these factors directly affect the volume of liquid in a tank.

The temperature is important because petroleum liquids expand—take up more volume—when warm, and contract—take up less volume—when cold. How much they expand and contract depends not only on their temperature but also on their density. For example, crude oil does not expand or contract as much as kerosene, given the same deviation in temperature, because crude oil is denser than kerosene. To measure how much liquid a tank contains, then, you must measure the temperature and the density of the liquid.

You must also determine the amount of suspended S&W in the liquid. Suspended S&W is extraneous matter that remains trapped in the tank oil. It can be seen only after being physically or chemically separated from the oil. You must subtract the amount of *suspended* S&W from the liquid volume just as you do the amount of *settled* S&W. Producers and carriers agree to limit the total amount of suspended S&W to a certain percentage of the liquid because S&W can cause corrosion and problems in processing and transporting; may violate federal, state, or municipal regulations; and may not be acceptable to the pipeline transporting the liquid. If tests reveal a higher percentage than the one agreed on, the suspended S&W must be removed before the transfer takes place. Once you have measured the temperature, the density of the liquid in the tank, and the suspended S&W content, you can determine the net volume.

MEASURING THE TEMPERATURE OF A LIQUID

The temperature of a stored liquid may vary throughout its depth. Therefore, when you measure temperature, you must take several readings and average them in order to calculate the volume accurately. You take these temperature readings before or after you gauge the liquid level. API *MPMS*, chapter 7, recommends that, in tanks with liquid more than 10 feet deep, you take at least three temperature measurements: one in the middle of the upper third of the liquid; one in the middle of the liquid; and one in the middle of the bottom third of the liquid. Settled S&W can affect the temperature reading because it can take longer to cool than oil does. So, if you take a reading near the bottom of the liquid, be sure you know the height of the settled S&W and position the thermometer above it.

In tanks with liquid heights less than 10 feet deep, take at least one temperature measurement in the middle of the liquid (table 3.1).

Read the thermometer to the nearest 1°F each time you measure the temperature and record the reading. Average all required readings and round off to the nearest whole degree. Since the atmospheric temperature can make the tank shell warmer or cooler than the liquid in the tank, take temperature measurements at least 12 inches from the side of the shell, or as far from the tank shell as possible.

TABLE 3.1
Minimum Number of Temperature Measurements for Various Depths of Hydrocarbon Liquid in Storage, Lease, Ship, and Barge Tanks

Depth of Liquid	Minimum No. of Temperature Measurements	Measurement Levels
>10 ft (3.05 m)	3	Middle of upper, middle, and lower thirds
≤10 ft (3.05 m)	1	Middle of liquid

Notes:

1. For tanks with capacities less than 5,000 barrels, one temperature measurement at the middle of the liquid can be used. One temperature measurement at the middle of the liquid will also suffice in ship or barge tanks containing less than 5,000 barrels.

2. The temperature of a liquid in a storage tank or marine vessel can vary throughout its depth; therefore, when temperature differentials greater than 5°F (3°C) are found, an average temperature shall be obtained. This may be accomplished by taking temperatures at different levels that are equally spaced, averaging the readings, rounding the result to the nearest 1.0°F (0.5°C), and reporting the result as the average temperature for the entire volume. In some cases, such as when a tank has a nonuniform cross-sectional area, it may be necessary to calculate a volume-weighted average temperature. Temperatures may be reported to less than 1.0°F (0.5°C) if all parties involved agree and it does not conflict with legal requirements.

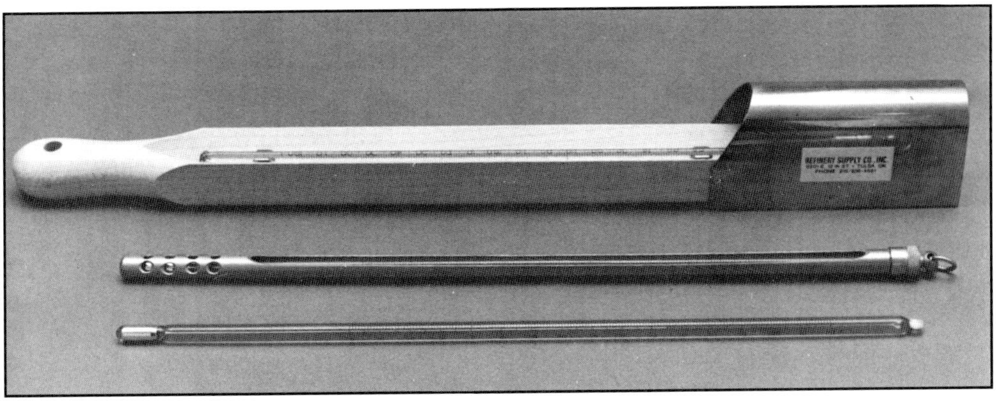

Figure 3.1. Portable thermometers

Thermometers

You may choose from a variety of thermometers to measure the temperature of a liquid, depending on the range of temperature of that liquid and the type of tank or vessel holding the liquid. Some thermometers are portable and are carried from site to site (fig. 3.1) while others are permanently mounted on the tank (fig. 3.2). Most thermometers reflect temperatures in at least 1°F increments. Some electronic thermometers, however, indicate temperature increments to the nearest 0.1°F.

All liquid in glass thermometers, whether portable or permanently installed, should be in excellent condition. The liquid column should not be separated, and the glass should have clear, readable etchings. If using a dial thermometer, meter-line thermometer, or permanently installed thermometer, make sure the glass is clear and free of cracks. All thermometers should be calibrated at least once a year using either an NIST thermometer or a verified calibrated thermometer traceable to an NIST thermometer. On permanently installed thermometers, the date of the last calibration should appear on the dial or scale face and calibration records must be kept.

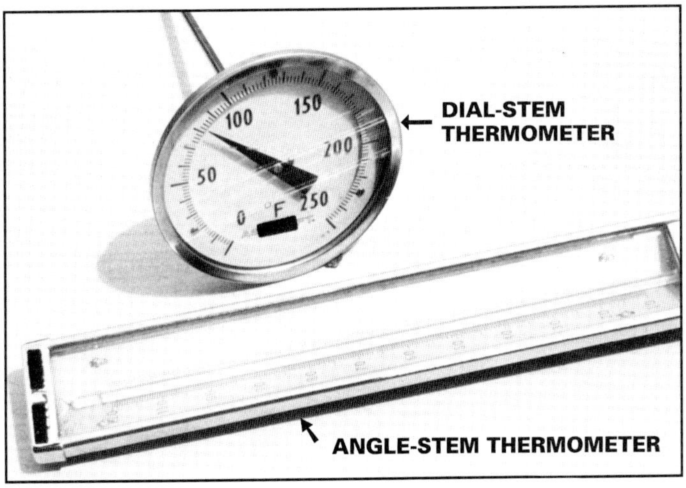

Figure 3.2. Permanently mounted thermometers

All thermometers require a certain amount of time to reach equilibrium with the liquid being measured. The amount depends on the API gravity at 60°F of the oil or product being measured and the type of thermometer being used. Be sure to check the recommended immersion times for each liquid and thermometer and keep the thermometer at the proper level for the specified time in order to obtain an accurate reading. These immersion times (for both stationary and "in-motion" thermometers) are stated in *MPMS,* chapter 7, section 1, "Temperature Determination."

Which thermometer you use depends on several factors. Table 3.2 reproduces a portion of table 1 in API

TABLE 3.2
Thermometer Assemblies and Temperature Levels for Tanks and Cargo Carriers

Type of Tank	Temperature Measurement Facilities	Thermometer Assembly	Measurement Level
	Stationary Tanks		
Fixed-roof	Roof hatch	Cup-case	See table 3.1
Floating-roof	Gauge hatch	Cup-case	See table 3.1
Variable vapor space	Gauge hatch	Cup-case	See table 3.1
Pressure	Vertical thermowells	Armored	See table 3.1
	Pressure lock	Cup-case	See table 3.1
	Horizontal separable wells	Angle-stem	3 for tank heights less than 30 ft (9.14 m); 4 for tank heights greater than 30 ft (9.14 m)
Horizontal-cylindrical			
Nonpressure	Gauge hatch	Cup-case	See table 3.1
Pressure	Vertical thermowells	Armored	See table 3.1
	Horizontal separable wells	Angle-stem	2, 1 at middle of tank and one 1 ft (0.3 m) above bottom
	Tank Cars and Tank Trucks		
Nonpressure	Dome hatch	Cup-case	Middle of liquid[a]
Pressure	Vertical thermowells	Armored	Middle of liquid[a]
	Ships and Barges		
Nonpressure	Deck hatch	Cup-case	See table 3.1
Pressure	Vertical thermowells	Armored	Middle of liquid[a]

Source: API, *MPMS,* chap. 7, table 1.

[a]In special cases, it may be desirable to obtain temperatures at more than one level and calculate a weighted average temperature. See table 3.1, note 2, for the correct method of averaging multiple temperatures.

MPMS, chapter 7, which recommends thermometers for use in different types of tanks. Table 3.3 reproduces table 2 in API *MPMS*, chapter 7. It lists certain thermometers, the range of each, its length, its degree of accuracy, and the measuring increments. No matter which thermometer you use, make sure that it is clean, that it is immersed in liquid at the specified level, and that it remains at that level long enough to reach equilibrium with the liquid.

TABLE 3.3
Thermometers

Type of Thermometer	ASTM Thermometer	Range	Length (in.)	Graduation	Accuracy
ASTM tank	58°F–80	–30°F–+120°F	12	1°F	±0.5°F
ASTM tank	97°F–80	0°F–120°F	12	1°F	±0.5°F
ASTM tank	59°F–80	0°F–180°F	12	1°F	±0.5°F
ASTM tank	98°F–80	60°F– 180°F	12	1°F	±0.5°F
ASTM tank	60°F–80	170°F– 5000°F	12	2°F	±1.0°F
Angle-stem	—	Suitable range	12[a]	1°F	±1.0°F
Tank[b]	—	20°F– 220°F	12	1°F	±0.5°F

Source: API, *MPMS*, chap. 7, table 2.

Notes: Except for the angle-stem thermometer, all of the thermometers listed are the total-immersion type.

[a]Length of graduated portion

[b]This thermometer does not have an ASTM designation, but it is commonly used for certain heated materials.

Portable Thermometer Assemblies

API recommends three types of portable thermometer assemblies: the cup-case assembly; the armored-case assembly; and the electronic assembly. Which assembly you use depends on company policy, manufacturer preferred, and availability.

CUP-CASE ASSEMBLY

The cup-case assembly is widely used because you can use it in nonpressure tanks, vessel compartments, tank cars and trucks, and stationary pressure tanks equipped with gauging locks. This assembly includes a liquid-in-glass thermometer clamped to a case of varnished hardwood or corrosion-resistant metal and a corrosion-resistant metal cup (fig. 3.3), with, in some instances, a flapper cover for holding liquid in the cup.

When using a cup-case thermometer, note whether the atmospheric temperature differs more than 20°F from the liquid temperature. If it does, you should immerse the thermometer two or more times in the upper portion of the liquid in the tank, filling and emptying—that is, flushing—the cup each time. This procedure will help the thermometer reach equilibrium more quickly. After flushing the cup, lower the assembly to the specified level (see table 3.1), and leave it for the correct amount of time (table 3.4), determined by the API gravity of the liquid and the temperature differential between the readings. Notice that table 3.4 reflects different immersion times, depending on whether the assembly is *in-motion* or *stationary*. *In-motion* refers to raising and

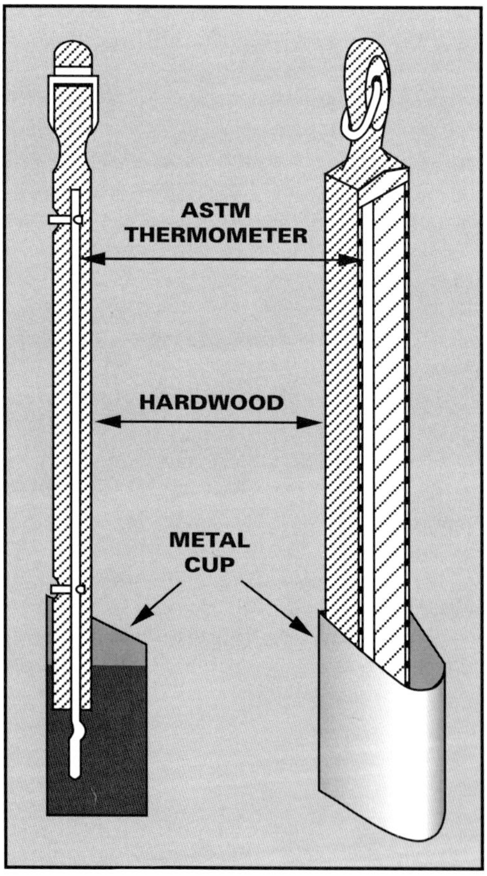

Figure 3.3. Typical cup-case assembly

lowering the assembly through a distance of 1 foot above and below the specified level to help the thermometer reach equilibrium more quickly. For example, if you lower the thermometer to the middle of the liquid to obtain a temperature measurement and that depth is 20 feet, raising and lowering the thermometer from 19 to 21 feet for a period of time determined by the liquid's API gravity will shorten the minimum immersion time.

After the required immersion time has elapsed, raise the assembly making sure that the cup remains full of liquid and that the base of the thermometer is submerged in liquid. Keep the thermometer below the edge of the hatch to prevent wind and other atmospheric conditions from affecting the reading. The mercury column should be stable; if it isn't, repeat the immersion procedure. Once the column is stable, immediately read and record the temperature.

TABLE 3.4
Recommended Immersion Times for Woodback Cup-Case Assembly

	Recommended Immersion Times (min)[a]	
API Gravity at 60°F	**In-Motion**	**Stationary**
Temperature differential greater than 5°F		
>50	5	10
40–49	5	15
30–39	12	25
20–29	20	45
<20	45	80
Temperature differential less than 5°F		
>50	5	10
40–49	5	15
30–39	12	20
20–29	20	35
<20	35	60

Notes:

[a]These immersion times were established based on the test procedure outlined in API, *MPMS*, chap. 7, Appendix A. Failure to use these recommended times may result in incorrect temperature readings.

1. The woodback cup-case assembly can be used in either an in-motion or a stationary mode. In-motion is defined as raising and lowering the assembly 1 ft (0.3 m) above and below the desired depth.

2. Cup-case assemblies made of other materials will have different immersion times. Immersion times should be established by testing and all parties involved should agree on the immersion times.

3. If additional mass, such as a weight to cause the woodback cup-case assembly to sink, is placed in the liquid near the thermometer, the immersion time of the assembly will be longer than those listed here. Immersion times should be established by testing and all parties involved should agree on the times.

The cup-case thermometer assembly can be used in pressure tanks with a pressure lock mounted on the top of the tank (fig. 3.4). When measuring the temperature of the liquid in these tanks, you first close the shutoff valve at the base of the lock to prevent tank vapors from escaping. Then you open the pressure-lock window to reach the tape and bob inside. Remove the bob and replace it with a thermometer assembly. Close the window and slowly open the shutoff valve. Using the external reel crank, lower the assembly to the specified immersion level. Allow enough time for the assembly to stabilize. Then, raise it into the pressure lock and read the temperature through the window. When you have completed the temperature measurement procedure, close the shutoff valve, remove the thermometer assembly, and replace the bob. Close the pressure-lock window and slowly reopen the valve.

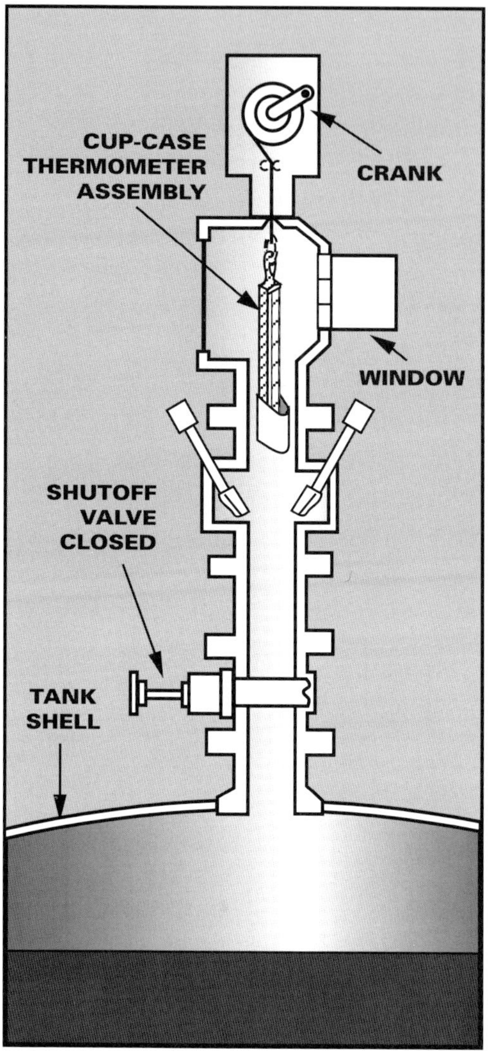

Figure 3.4. Cup-case thermometer assembly mounted in a pressure lock on top of a pressure tank

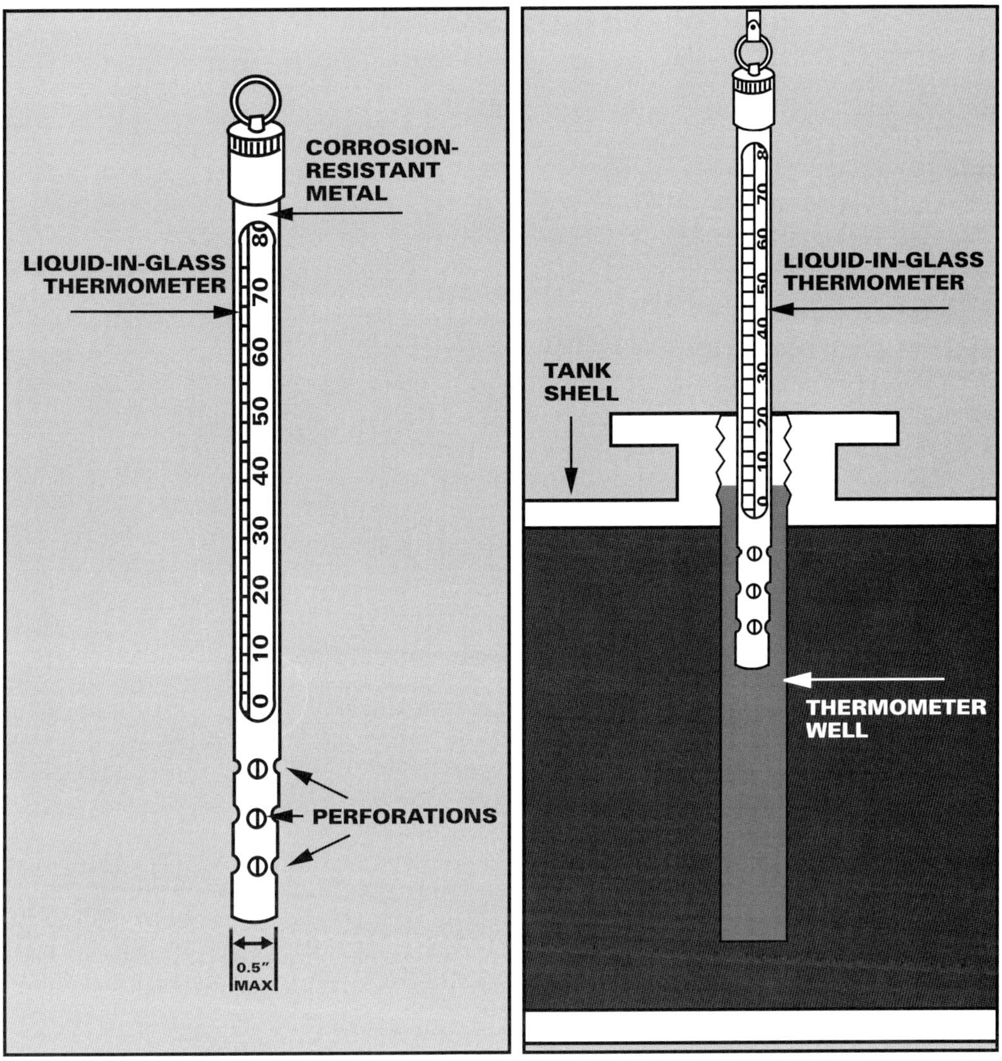

Figure 3.5. Typical armored-case assembly

ARMORED-CASE ASSEMBLY

The armored-case assembly consists of a liquid-in-glass thermometer enclosed by a corrosion-resistant, protective metal case. The lower portion of the case is perforated (fig. 3.5). Because of its slenderness, this assembly is used primarily in pressure tanks with vertical thermometer wells. A low-volatility oil, like kerosene, fills the well and helps transfer temperature from the liquid in the tank to the thermometer. This assembly is lowered to the specified level in the well and left there for at least 5 minutes. Liquid flows through the perforations in the case and surrounds the thermometer. After the required time has elapsed, the thermometer is raised to the surface. Since this assembly does not have a cup at its base, the perforated end must be kept submerged in liquid in the well, immediately read, and the temperature recorded.

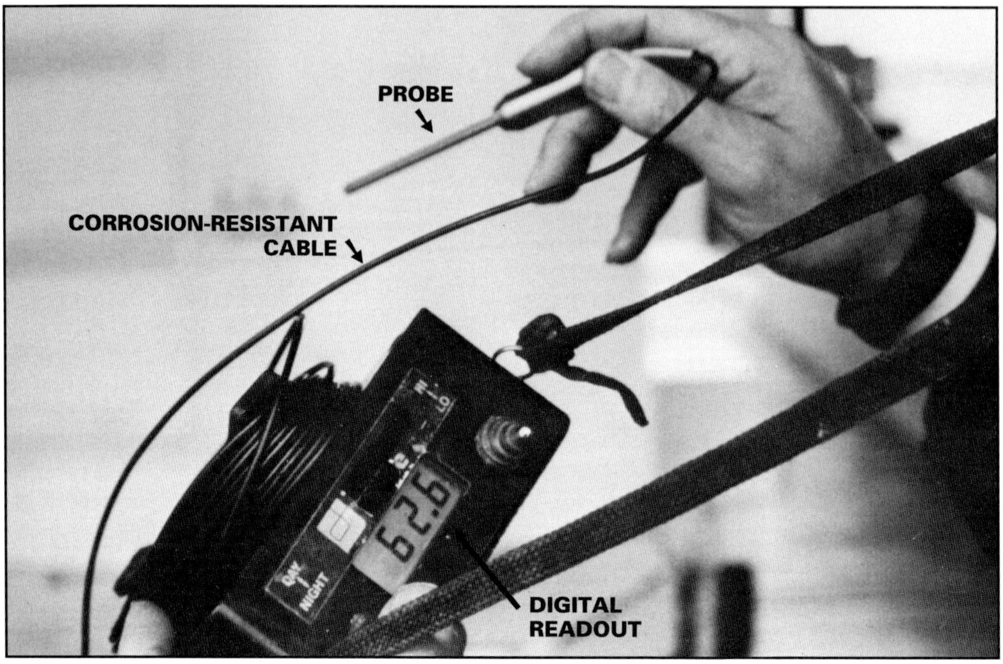

Figure 3.6. Typical electronic thermometer assembly

ELECTRONIC ASSEMBLY

Some portable thermometer assemblies are electronic. They consist of a probe, a flexible corrosion-resistant cable, and a meter housing with a digital readout (fig. 3.6). If you are using an electronic thermometer, you must calibrate it daily or before each use to ensure accuracy. You may use a cup-case assembly of known accuracy to calibrate an electronic thermometer.

When using an electronic thermometer, you lower the probe to the specified level. At this level, raise and lower the probe through a distance of about 1 foot for the specified time to help it reach temperature equilibrium. The temperature of the liquid is transmitted electronically to the main housing on the surface. Immediately read the display and record the temperature.

Permanent, or Fixed, Thermometer Assemblies

Some thermometer assemblies are permanently attached to the tank or vessel. Two commonly used types are the angle-stem assembly (fig. 3.7) and the dial-type assembly (fig. 3.8). The stem of the angle-stem assembly may have an angle of 90° or greater, depending on the contour of the tank shell. Angling the stem allows you to read the temperature of the liquid easily. The glass stem is protected with a light metal tube.

The dial-type assemblies may be either mercury or bimetal with a straight stem. The dial is attached to the stem. You can easily read the temperature of the liquid by reading the dial.

Both these assemblies are equipped with a threaded coupling that joins to a well or a socket permanently mounted in or on the tank. The stems of both these thermometers should be at least 3 feet long, unless they are used in a vertical tank of less than

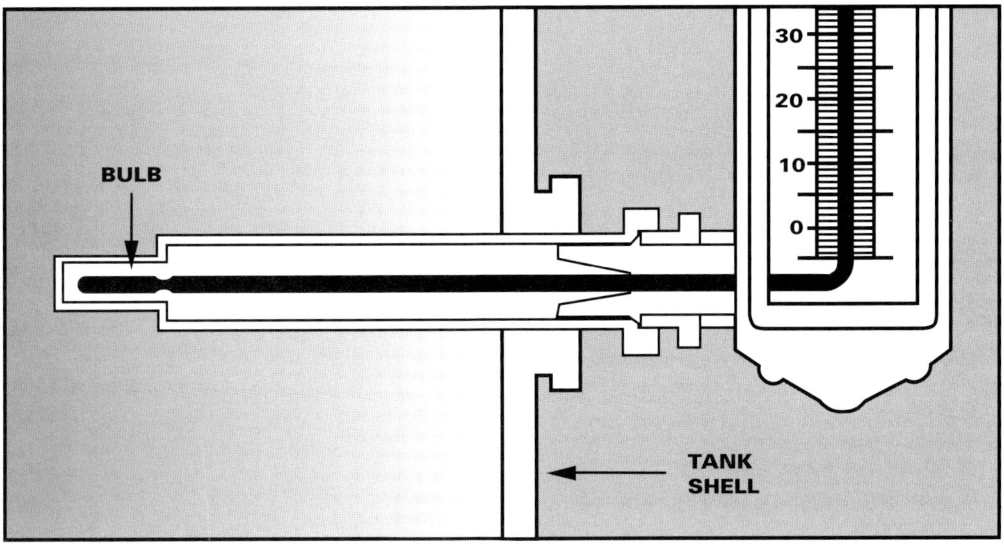

Figure 3.7. Typical angle-stem assembly

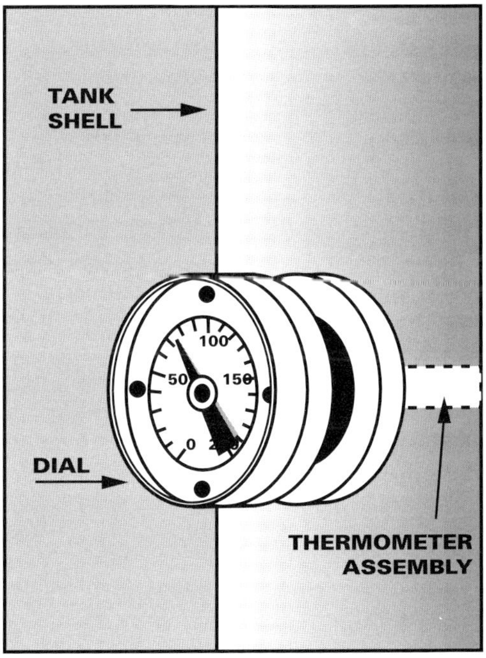

Figure 3.8. Dial-type assembly

5,000-barrel capacity. In these vertical tanks the stem may be a minimum of 12 inches long.

The stem of any permanently mounted thermometer assembly must be below the surface of the liquid. If it is not, the temperature reading will be inaccurate. The temperature-sensitive portion of both assemblies should not be more than 2.5 inches long. These thermometers must also not interfere with floating roofs, mixers, or any other moving parts in a tank.

MEASURING THE DENSITY OF A LIQUID

To measure the density of liquids in the field, you use an instrument called a hydrometer. The hydrometer is a graduated cylinder with a weight in the bottom.

Just as there are different scales for measuring temperature, such as Celsius or Fahrenheit, there are different scales for measuring density. The hydrometer scale can be graduated in *relative density* (specific gravity) or in °*API*. In the field, the term *API gravity* is almost universally used when referring to density and is expressed as °API. Just remember that relative density (specific gravity) and API gravity are different measures of the same thing—the density of the liquid.

Regardless of scale, the hydrometer is designed to float vertically in a sample of the liquid petroleum. How buoyant the hydrometer is depends on the relative density, or API gravity of the liquid in which it is floating. For example, a hydrometer will float, or sit, high up in a relatively dense liquid like crude oil, but will sink to a much lower position in a low-density liquid like kerosene.

Materials Needed for Measuring Density

To obtain an accurate reading of the relative density or API gravity of a liquid, you need four things: a clear glass hydrometer graduated in units of relative density or °API; a thermometer; a liquid sample from the tank that is representative of the entire contents; and a clear glass or metal cylinder for holding the sample. The hydrometer graduations must conform to ASTM (American Society of Testing and Materials) specifications or those listed in table 1 of API *MPMS*, chapter 9, section 1, "Density Determination." The thermometer must conform to ASTM specifications.

Because the sample must be representative of the entire amount of liquid in the tank, you should withdraw it following the procedures described in chapters 4 and 5 of this book or those described in API *MPMS*, chapter 8, "Manual Sampling of Petroleum and Petroleum Products." When measuring relative density or API gravity, the sample should be held in a cylinder with an inside diameter (ID) at least 25 millimetres greater than the outside diameter (OD) of the hydrometer. The cylinder should also be tall enough for the hydrometer to float in the sample with at least a 25-millimetre clearance between the bottom of the hydrometer and the bottom of the cylinder (fig. 3.9). A container this size ensures that the hydrometer floats freely without touching the sides or the bottom of the container.

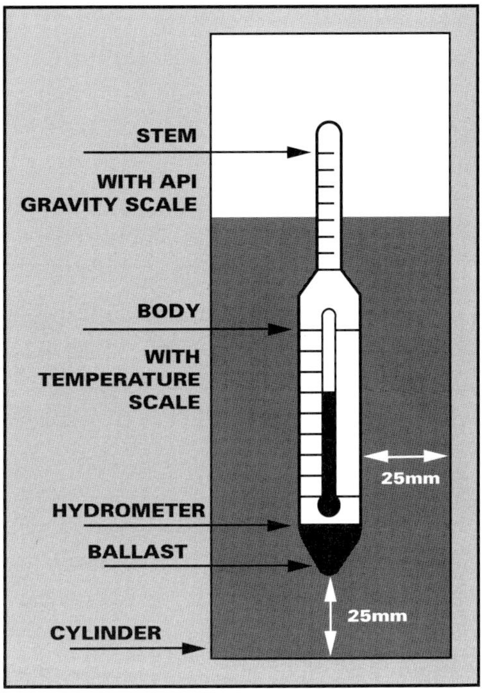

Figure 3.9. There must be at least 25 mm clearance between sides of hydrometer and cylinder and at least 25 mm clearance between bottom of cylinder and bottom of hydrometer.

Since relative density and API gravity are measures of the mass per unit volume (the density) of a substance at a certain temperature, you must accurately determine the temperature when measuring the density of the sample. Many hydrometers, therefore, are combined with a thermometer into one instrument called a thermohydrometer. If you are using a thermohydrometer, you can record the temperature and the relative density or °API at the same time. If you are not using this particular instrument, you must measure the temperature and the relative density or °API separately.

Procedures for Determining Density

THERMOHYDROMETER

Gently lower the thermohydrometer into the sample until it floats freely without touching the sides or the bottom of the cylinder. Do not drop the instrument or force it down into the liquid and let it bob back up. Any liquid on the stem above the point at which the instrument floats will weight the stem and give an erroneous reading. Depress the thermohydrometer about two scale divisions and then release it. Do not depress it more than two divisions or the weight of the liquid on the instrument stem will affect the reading. (The weight of two divisions is not enough to distort readings.) The thermohydrometer should immediately return to its previous position. If it doesn't, the liquid is too thick to test and probably needs to be heated. With low-viscosity liquids, give the thermohydrometer a slight spin after depressing it to help bring it to rest floating freely away from the walls of the cylinder. Allow sufficient time, usually 5 minutes, for the thermohydrometer to come to rest and for all the air bubbles to come to the surface. You may need to wait longer for more viscous samples.

When you are ready to read the instrument, notice that the liquid creeps upward where it comes in contact with the thermohydrometer stem. This creeping movement is caused by surface tension. The small amount of liquid that creeps up on the stem is called the meniscus (fig. 3.10). The correct hydrometer reading is that point on the scale at which the principal surface of the liquid cuts the scale. If the scale is in °API, read to the nearest 0.1°API at the top of the meniscus for opaque liquids (e.g., crude oil), then subtract one scale division. Record the remainder. For clear liquids (e.g., gasoline), read the bottom of the meniscus (the principal surface of the liquid) and record the reading. If the scale is in units of relative density, read it to the nearest 0.001 at the top of the meniscus for opaque liquids, subtract one scale division, and record the remainder; or read and record to the nearest 0.001 at the bottom of the liquid for clear liquids. Next, withdraw the instrument just enough to expose the thermometer scale without pulling the tip of the thermohydrometer out of the oil. Read and record the observed temperature to the nearest whole degree Fahrenheit.

Try to shelter the thermohydrometer from the wind and other elements as much as possible, to make sure that the sample temperature does not change more than 2°C (5°F) during the test.

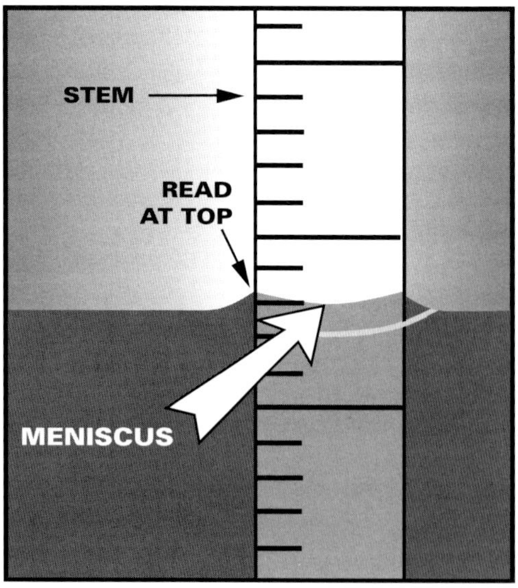

Figure 3.10. Read liquid at the top of the meniscus.

HYDROMETER

If the hydrometer and the thermometer are separate instruments, the measurement procedure is somewhat different. Gently lower the hydrometer into the sample. Again, make sure that you do not push the stem down into the liquid and get liquid on the stem above the level at which the instrument floats. While the hydrometer is floating, use the thermometer to stir the sample continuously. Take care that the mercury column on the thermometer is kept immersed while stirring and that your stirring action does not wet the hydrometer stem above the level at which it floats. As soon as a steady reading is obtained, record the temperature of the sample to the nearest 1°F and then remove the thermometer.

Once you have recorded the temperature of the sample, read and record the relative density or °API as you do with a thermohydrometer. Depress the hydrometer about two scale divisions and then release it, making sure the instrument immediately returns to its previous position. (Remember that the weight of two scale divisions does not distort the reading.) In low-viscosity liquids, give the hydrometer a slight spin after depressing it and allow sufficient time for all air bubbles to come to the surface. If the scale is in °API, read to the nearest 0.1°API at the top of the meniscus for opaque liquids (e.g., crude oil), then subtract one scale division and record the remainder. For clear liquids (e.g., gasoline), read the bottom of the meniscus (the principal surface of the liquid) and record the reading. If the scale is in units of relative density, read it to the nearest 0.001 at the top of the meniscus for opaque liquids, subtract one scale division, and record the remainder; or read and record to the nearest 0.001 at the bottom of the liquid for clear liquids.

MEASURING THE SUSPENDED S&W CONTENT

The third measurement you will make is to determine the amount of suspended S&W in the oil or product. One of the most commonly used procedures involves obtaining a representative sample of oil or product from the tank, putting a specific amount of this sample in a glass tube, adding solvent, adding demulsifier if necessary, heating the liquid in the tube to a specified temperature, and then rotating the tube in a centrifuge for a specific amount of time at a specific speed. (The amount, temperature, and time are found in API *MPMS*, chapter 10, section 4.) The centrifugal force created in the centrifuge forces the heavier S&W to the bottom of the tube. When you remove the glass tube from the centrifuge, you can read the amount of sediment and water that has been forced to tube bottom. The percentage of S&W in the tube is the percentage of S&W entrained in the oil.

Materials Needed to Determine Suspended S&W

To determine suspended S&W content, you need a representative sample of the liquid in the tank, a preheater, a thermometer, graduated cone-shaped tubes made of annealed glass, a solvent, and a centrifuge. If you are dealing with stable emulsions, you will also need a demulsifier. A solvent is used when testing because water and sediment do not fall out of oils very easily. A solvent thins the oil, that is, makes it less viscous, so that the S&W more easily falls to the bottom of the tube.

API *MPMS*, chapter 10, section 4, "Determination of Sediment and Water in Crude Oil by the Centrifuge Field Method," lists the recommended solvents as well as fire, safety, and health considerations related to their use. Some solvents must be water-saturated, since water in the oil sample can go into solution with these solvents if they are not. API *MPMS*, chapter 10, section 3, "Determination of Sediment and Water in Crude Oil by the Centrifuge Lab Method," also includes procedures for saturating a solvent with water if required.

Sometimes a solvent is not the only chemical needed when testing. A demulsifier may also be required. The demulsifier chemically breaks an emulsion, allowing the water droplets suspended in the oil to merge and fall to the bottom of the container. When a demulsifier is used, it should always be in the form of a demulsifier-solvent stock solution or be premixed with the solvent to be used in the test. A number of demulsifiers may be used; just be sure you follow the manufacturer's recommendations regarding health and safety factors.

You should use one of two types of centrifuge tubes recommended by API *MPMS*, chapter 10, sections 3 and 4. These tubes are cone-shaped and made of annealed glass. Both tubes are designed so that they can be closed with a stopper. They are either 203 millimetres (8 inches) or 167 millimetres (6 inches) long and are graduated in millilitres, or parts (fig. 3.11).

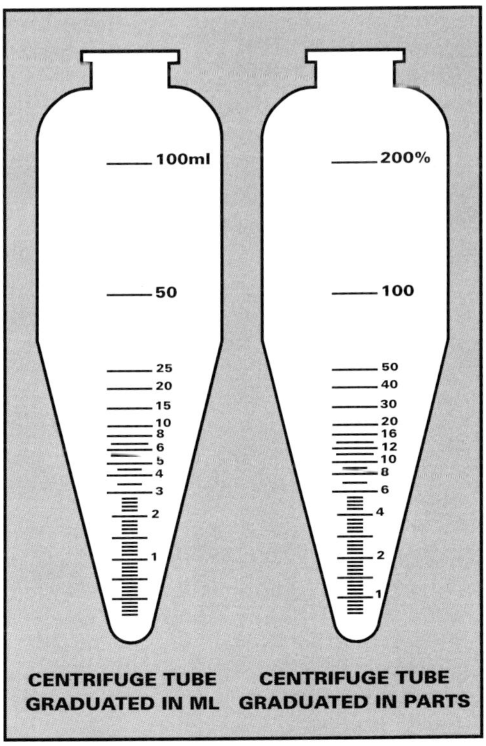

Figure 3.11. Centrifuge tubes

The difference between the graduations on these tubes can be confusing when you are determining the percentage of S&W in the liquid. One millilitre equals 2 parts, so, 25 millilitres in one tube equals 50 parts in the other. One part *also* equals 1% when reading the percentage of S&W in the bottom of the tube. If you are using the tube graduated in parts, converting to percentages is an even exchange—1 part equals 1%. If, however, you are using the tube graduated in millilitres, you must double the number before recording the percentage of S&W—0.5 millilitre × 2 = 1.0%.

API recommends that the thermometer you use have graduations at intervals of 1°C (2°F) and be accurate to ± 1°C (2°F).

You should use a preheater to heat the sample before you centrifuge it. Heating the sample makes it less viscous so that the S&W can more easily be centrifuged out of the liquid and collected at the bottom of the centrifuge tube. According to API, the preheater should be either a metal block or a liquid bath of sufficient depth to permit immersion of the centrifuge tube in a vertical position to the 100-millimetre (200-part mark). The preheater should also be capable of heating the sample to a minimum of 60°C ±3°C (140°F ±5°F).

The centrifuge you use should meet the requirements listed in API *MPMS*, chapter 10, section 4. In general, the centrifuge must be capable of spinning two or more centrifuge tubes fast enough to create a relative centrifugal force of at least 500. Table 3.5 lists the applicable rotation speeds for centrifuges of various swing diameters. All centrifuge parts have to be constructed to withstand the maximum centrifugal force generated by the unit. The unit must also be enclosed by a metal shield or case strong enough to contain flying debris if a tube breaks or if the centrifuge malfunctions. The centrifuge should have a heater so that it can maintain the sample above the minimum allowable temperature of 52°C (125°F) when centrifuging is completed.

TABLE 3.5
Rotation Speeds to Produce a Relative Centrifugal Force of 500 for Centrifuges of Various Diameters of Swing

Diameter of Swing[a]		Rotation Speed (rpm)
Millimetres	Inches	
305	12	1,710
330	13	1,640
356	14	1,580
381	15	1,530
406	16	1,480
432	17	1,440
457	18	1,400
483	19	1,360
508	20	1,325
533	21	1,290
559	22	1,260
584	23	1,240
610	24	1,210

Source: API, *MPMS*, chap. 10, section 4.

[a]Measured between the tips of opposite tubes when the tubes are in rotating position.

Procedure to Determine S&W Content

Fill two centrifuge tubes to exactly the 50-millilitre (100-part) mark with a sample taken from the container in which the sample was collected or from the sampling device. The sampling device may be a thief, a bottle, a beaker, or a LACT sample container. Read the top of the meniscus to determine the 50-millilitre (100-part) mark. Fill each tube with the solvent to exactly the 100-millilitre (200-part) mark. Again, read the top of the meniscus to determine the 100-millilitre (200-part) mark. If experience indicates that a demulsifying agent is required and one has not already been added to the solvent, add to each tube a drop or two of demulsifier-solvent stock solution that has previously been determined to be satisfactory for the crude oil under test.

Next, stopper each tube tightly and invert the tubes a minimum of ten times to ensure that the oil and the solvent are uniformly mixed. Remember to invert the tubes below eye level in case the stopper blows out. Vapor pressure from the agitated hydrocarbons tends to increase as the temperature of the liquid increases. For example, the vapor pressures of hydrocarbons at 60°C (140°F) are about double those at 40°C (104°F), so be careful—stopper blowout is possible. In cases where the crude oil is particularly viscous and it is difficult to mix the solvent with the oil, the solvent may be added to the centrifuge tube prior to the oil to make mixing easier. Be especially careful to fill the centrifuge tubes to exactly the 50-millilitre (100-part) mark with solvent and exactly to the 100-millilitre (200-part) mark with the sample. Remember to read at the top of the meniscus.

After mixing the solvent and the sample in the appropriate manner, loosen the stoppers to prevent pressure buildup during heating and immerse the tubes to the 100-millilitre (200-part) mark in a preheater. Heat the contents to 60°C ± 3°C (140°F ± 5°F). Next, secure the stoppers and again invert the tubes ten times to ensure uniform mixing of the oil and solvent. (Remember to keep the tubes below eye level.) Place the tubes in the trunnion cups on opposite sides of the centrifuge to establish balance. Retighten the stoppers and spin for at least 5 minutes at a minimum relative centrifugal force of 500.

Immediately after the centrifuge head comes to rest, verify the temperature of the samples in the tubes, but do not disturb the oil-water interface at the bottom of the tube with the thermometer. The test is invalid if the final temperature after centrifuging is below 52°C (125°F) and must be repeated with a fresh sample.

Read and write down the combined volume of S&W at the bottom of each tube. If the amount of S&W is 0.10 millilitre (0.20% for 200-part tubes) or below, and if the difference between the two readings is greater than one subdivision on the tube, or 0.025 millilitre (0.05% for 200-part tubes), the readings are inadmissible and the test must be repeated.

If the tubes are graduated in parts, write down the average of the values read directly from the two tubes. When finding the average, calculate it to the third decimal place. This average represents the percentage of S&W present in the liquid. Remember, each part on this tube equals 1%.

If the tubes are graduated in millilitres, record the sum of the final volumes of S&W from each tube. Remember, 1 millilitre equals 2 parts, and 1 part equals 1%. So this sum is the percentage of S&W present in the liquid.

After the initial readings are made and written down, reheat both tubes to 60°C ± 3°C (140°F ± 5°F), return the tubes without agitation to the centrifuge and spin again for 5 minutes. Repeat this operation until two consecutive consistent readings are obtained for each tube. Finally, record the results in the tally book.

For the test to be considered valid, you must be able to see a clear interface between the oil and the separated water. No layering, such as from an emulsion, should be present immediately above the oil-water interface. If an emulsion layer is present, you may need to perform one or more of the following remedies: shake the mixture between whirlings in the centrifuge just enough to disperse the emulsion; use a different or an increased amount of demulsifying agent; or use a different or increased amount of solvent. After a satisfactory procedure for a particular type of oil or product has been worked out, it will normally be suitable for all samples of the same type.

CALCULATING NET VOLUME

Once you have determined the temperature, API gravity, and S&W content of a liquid stored in a tank, you can calculate the net volume. Suppose the tank you gauged has 32 feet of liquid in it. After checking the gauge tables computed for this tank, you find that 32 feet equals 55,000 barrels of liquid. Fifty-five thousand barrels is the *gross* volume. Next, you take the measurement of settled S&W, in this case, 2 feet, and subtract that amount from the gross volume. Checking your gauge tables again, you see that 2 feet equals 1,720 barrels. Subtract 1,720 from 55,000 to determine the gross amount of sellable liquid—53,280 barrels. Next, take the measurements of API gravity and temperature that you obtained from the representative sample and refer to the available tables to determine the API gravity of the liquid adjusted to 60°F. Table 3.6 reproduces a portion of the Petroleum Measurement Tables, Volume I, Table 5A, Generalized Crude Oils. Suppose your measurements determined that the API gravity of the sample was 23.5° at an observed temperature of 84°F. To find the adjusted API gravity at 60°F, go down the column marked 23.5° API until you are opposite the line for 84°F. The adjusted API gravity at 60°F is 22.0°API.

TABLE 3.6
Generalized Crude Oils, API Gravity Correction to 60°F

	API Gravity at Observed Temperature										
Temp F	20.0	20.5	21.0	21.5	22.0	22.5	23.0	23.5	24.0	24.5	25.0	**Temp F**
	Corresponding API Gravity at 60°F											
75.0	19.1	19.6	20.1	20.6	21.1	21.6	22.1	22.6	23.1	23.6	24.1	75.0
75.5	19.1	19.6	20.1	20.6	21.1	21.6	22.1	22.6	23.0	23.5	24.0	75.5
76.0	19.1	19.6	20.1	20.6	21.1	21.6	22.1	22.6	23.1	23.6	24.1	76.0
76.5	19.1	19.6	20.0	20.5	21.0	21.5	22.0	22.5	23.0	23.5	24.0	76.5
77.0	19.0	19.5	20.0	20.5	21.0	21.5	22.0	22.5	23.0	23.4	23.9	77.0
77.5	19.0	19.5	20.0	20.5	21.0	21.5	21.9	22.4	22.9	23.4	23.9	77.5
78.0	19.0	19.5	20.0	20.4	20.9	21.4	21.9	22.4	22.9	23.4	23.9	78.0
78.5	19.0	19.4	19.9	20.4	20.9	21.4	21.9	22.4	22.9	23.4	23.8	78.5
79.0	18.9	19.4	19.9	20.4	20.9	21.4	21.9	22.3	22.8	23.3	23.8	79.0
79.5	18.9	19.4	19.9	20.4	20.8	21.3	21.8	22.3	22.8	23.3	23.8	79.5
80.0	18.9	19.4	19.8	20.3	20.8	21.3	21.8	22.3	22.8	23.3	23.7	80.0
80.5	18.8	19.3	19.8	20.3	20.8	21.3	21.8	22.3	22.7	23.2	23.7	80.5
81.0	18.8	19.3	19.8	20.3	20.8	21.3	21.7	22.2	22.7	23.2	23.7	81.0
81.5	18.8	19.3	19.8	20.2	20.7	21.2	21.7	22.2	22.7	23.2	23.7	81.5
82.0	18.8	19.2	19.7	20.2	20.7	21.2	21.7	22.2	22.7	23.1	23.6	82.0
82.5	18.7	19.2	19.7	20.2	20.7	21.2	21.6	22.1	22.6	23.1	23.6	82.5
83.0	18.7	19.2	19.7	20.2	20.6	21.1	21.6	22.1	22.6	23.1	23.6	83.0
83.5	18.7	19.2	19.6	20.1	20.6	21.1	21.6	22.1	22.6	23.0	23.5	83.5
84.0	18.6	19.1	19.6	20.1	20.6	21.1	21.6	22.0	22.5	23.0	23.5	84.0
84.5	18.6	19.1	19.6	20.1	20.6	21.0	21.5	22.0	22.5	23.0	23.5	84.5
85.0	18.6	19.1	19.6	20.0	20.5	21.0	21.5	22.0	22.5	23.0	23.4	85.0
85.5	18.6	19.0	19.5	20.0	20.5	21.0	21.5	22.0	22.4	22.9	23.4	85.5
86.0	18.5	19.0	19.5	20.0	20.5	21.0	21.4	21.9	22.4	22.9	23.4	86.0
86.5	18.5	19.0	19.5	20.0	20.4	20.9	21.4	21.9	22.4	22.9	23.4	86.5
87.0	18.5	19.0	19.4	19.9	20.4	20.9	21.4	21.9	22.4	22.8	23.3	87.0
87.5	18.4	18.9	19.4	19.9	20.4	20.9	21.4	21.8	22.3	22.8	23.3	87.5
88.0	18.4	18.9	19.4	19.9	20.4	20.8	21.3	21.8	22.3	22.8	23.3	88.0
88.5	18.4	18.9	19.4	19.8	20.3	20.8	21.3	21.8	22.3	22.7	23.2	88.5
89.0	18.4	18.8	19.3	19.8	20.3	20.8	21.3	21.7	22.2	22.7	23.2	89.0
89.5	18.3	18.8	19.3	19.8	20.3	20.8	21.2	21.7	22.2	22.7	23.2	89.5
90.0	18.3	18.8	19.3	19.8	20.2	20.7	21.2	21.7	22.2	22.7	23.1	90.0

Source: API Standard 2540, Petroleum Measurement Tables, Volume Correction Factors, vol. I, table 5A.
Note: API gravity = 20.0–25.0.

Now that you have the adjusted API gravity of the liquid, look at table 3.7. This table gives you a factor for adjusting the total liquid volume to what it would be at 60°F. Suppose that you determined that the average temperature of the liquid in the tank was 76°F. Table 3.7 lists temperatures down the left side and adjusted API gravity across the top. The adjusted gravity was 22.0°API at 60°F. The average temperature in the tank was 76°F. Going down the column marked 22.0°API until you are opposite 76°F, you find the factor for correcting the volume to 60°F—0.9936. The volume of liquid less the settled S&W was 53,280 barrels. Take this figure and multiply it by 0.9936. The product, 52,939.01 barrels, is the volume adjusted to 60°F.

Your percentage of suspended S&W was 0.005. That means half of 1% of the liquid is suspended S&W. Half of 1% of 52,939.01 is 264.69 barrels. Since suspended S&W is extraneous matter, subtract 264.69 from 52,939.01. The result is the net volume of sellable liquid in the tank—52,674.32 barrels of liquid at 22.0°API adjusted to 60°F.

In tabular form—

Step 1

55,000	gross volume
−1,720	barrels tank bottoms
53,280	barrels (gross observed volume)

Step 2

53,280	barrels
× .9936	factor adjusting for temperature and gravity
52,939.01	barrels (gross standard volume)

Step 3

52,939.01	barrels
× .005	% suspended S&W
264.69	barrels suspended S&W

Step 4

52,939.01	barrels
−264.69	barrels suspended S&W
52,674.32	barrels sellable liquid, 22.0° API adjusted to 60°F (net standard volume)

TABLE 3.7

Generalized Crude Oils, API Volume Correction to 60°F

	API Gravity at 60°F											
Temp F	20.0	20.5	21.0	21.5	22.0	22.5	23.0	23.5	24.0	24.5	25.0	**Temp F**
	Factor for Correcting Volume to 60°F											
75.0	0.9941	0.9941	0.9940	0.9940	0.9940	0.9939	0.9939	0.9938	0.9938	0.9938	0.9937	75.0
75.5	0.9939	0.9939	0.9938	0.9938	0.9938	0.9937	0.9937	0.9936	0.9936	0.9935	0.9935	75.5
76.0	0.9937	0.9937	0.9936	0.9936	0.9936	0.9935	0.9935	0.9934	0.9934	0.9933	0.9933	76.0
76.5	0.9935	0.9935	0.9934	0.9934	0.9934	0.9933	0.9933	0.9932	0.9932	0.9931	0.9931	76.5
77.0	0.9933	0.9933	0.9932	0.9932	0.9931	0.9931	0.9931	0.9930	0.9930	0.9929	0.9929	77.0
77.5	0.9931	0.9931	0.9930	0.9930	0.9929	0.9929	0.9929	0.9928	0.9928	0.9927	0.9927	77.5
78.0	0.9929	0.9929	0.9928	0.9928	0.9927	0.9927	0.9926	0.9926	0.9926	0.9925	0.9925	78.0
78.5	0.9927	0.9927	0.9926	0.9926	0.9925	0.9925	0.9924	0.9924	0.9923	0.9923	0.9922	78.5
79.0	0.9925	0.9925	0.9924	0.9924	0.9923	0.9923	0.9922	0.9922	0.9921	0.9921	0.9920	79.0
79.5	0.9923	0.9923	0.9922	0.9922	0.9921	0.9921	0.9920	0.9920	0.9919	0.9919	0.9918	79.5
80.0	0.9921	0.9921	0.9920	0.9920	0.9919	0.9919	0.9918	0.9918	0.9917	0.9917	0.9916	80.0
80.5	0.9919	0.9919	0.9918	0.9918	0.9917	0.9917	0.9916	0.9916	0.9915	0.9915	0.9914	80.5
81.0	0.9918	0.9917	0.9916	0.9916	0.9915	0.9915	0.9914	0.9914	0.9913	0.9913	0.9912	81.0
81.5	0.9916	0.9915	0.9914	0.9914	0.9913	0.9913	0.9912	0.9912	0.9911	0.9910	0.9910	81.5
82.0	0.9914	0.9913	0.9912	0.9912	0.9911	0.9911	0.9910	0.9910	0.9909	0.9909	0.9908	82.0
82.5	0.9912	0.9911	0.9910	0.9910	0.9909	0.9909	0.9908	0.9907	0.9907	0.9906	0.9906	82.5
83.0	0.9910	0.9909	0.9908	0.9908	0.9907	0.9907	0.9906	0.9906	0.9905	0.9904	0.9904	83.0
83.5	0.9908	0.9907	0.9906	0.9906	0.9905	0.9905	0.9904	0.9903	0.9903	0.9902	0.9901	83.5
84.0	0.9906	0.9905	0.9904	0.9904	0.9903	0.9903	0.9902	0.9901	0.9901	0.9900	0.9899	84.0
84.5	0.9904	0.9903	0.9902	0.9902	0.9901	0.9901	0.9900	0.9899	0.9899	0.9898	0.9897	84.5
85.0	0.9902	0.9901	0.9900	0.9900	0.9899	0.9899	0.9898	0.9897	0.9897	0.9896	0.9895	85.0
85.5	0.9900	0.9899	0.9898	0.9898	0.9897	0.9897	0.9896	0.9896	0.9895	0.9894	0.9893	85.5
86.0	0.9898	0.9897	0.9896	0.9896	0.9895	0.9894	0.9894	0.9893	0.9892	0.9892	0.9891	86.0
86.5	0.9896	0.9895	0.9894	0.9894	0.9893	0.9892	0.9892	0.9891	0.9890	0.9890	0.9889	86.5
87.0	0.9894	0.9893	0.9892	0.9892	0.9891	0.9890	0.9890	0.9889	0.9888	0.9887	0.9887	87.0
87.5	0.9892	0.9891	0.9890	0.9890	0.9889	0.9888	0.9888	0.9887	0.9886	0.9885	0.9885	87.5
88.0	0.9890	0.9889	0.9888	0.9888	0.9887	0.9886	0.9886	0.9885	0.9884	0.9883	0.9883	88.0
88.5	0.9888	0.9887	0.9886	0.9886	0.9885	0.9884	0.9883	0.9883	0.9982	0.9981	0.9980	88.5
89.0	0.9886	0.9885	0.9884	0.9884	0.9883	0.9982	0.9981	0.9981	0.9980	0.9979	0.9978	89.0
89.5	0.9884	0.9883	0.9982	0.9982	0.9981	0.9980	0.9979	0.9979	0.9978	0.9877	0.9876	89.5
90.0	0.9982	0.9981	0.9980	0.9980	0.9979	0.9978	0.9877	0.9877	0.9876	0.9875	0.9874	90.0

Source: API Standard 2540, Petroleum Measurement Tables, Volume Correction Factors, vol. I, table 6A.

Note: API gravity = 20.0–25.0

CONCLUSION

In the foregoing example, the net volume of oil in the tank was considerably different from the gross volume determined by gauging the tank. Whether this calculation is done for inventory or for custody transfer, it is important that it be accurate—small errors can translate into large amounts of money. Both parties involved in a custody transfer will measure the amount of oil or product. The two calculations must be the same. In cases in which they are not, both parties measure the oil or product again, often at the same time and with additional witnesses, until a satisfactory determination is made.

CHAPTER 4

Manual Sampling of Petroleum and Petroleum Products

INTRODUCTION

To the inexperienced, obtaining a sample of oil or product from a storage tank would not seem that difficult. Samples are relatively small, so why not just lower a suitable container into a tank, fill it, raise it to the surface and use the contents for whatever tests are needed?

If the liquids were anything but oil or oil products, this method might work. But oil and oil products are combinations of hydrocarbons and other substances. Storage tanks rarely contain homogeneous quantities of these liquids from top to bottom. The hydrocarbons and other materials tend to gravitate by weight into different strata throughout the tank. Just grabbing a sample from the top will not give you a good cross section of all of the vessel's contents—and obtaining a good cross section, or representative sample, is absolutely necessary. You use samples not only to test for API gravity and suspended sediment and water, but also to determine certain other physical and chemical characteristics that must be known before a custody transfer can take place or a price can be determined. For example, you use a sample to test the sulfur content of diesel fuels, the octane level of certain fuels, or to determine whether the product meets the buyer's or the government's specifications. If you do not have a sample that accurately represents the entire contents of the tank, then all tests on that sample will give you erroneous information.

Automatic sampling of pipeline systems is preferred. Automatic systems are discussed in detail in chapter 5 of this book and in API *MPMS,* chapter 8, section 2, "Automatic Sampling of Petroleum and Petroleum Products." In general, an automatic system provides a constant, systematic withdrawal of small amounts of crude oil or product as the liquid leaves the tank and enters the pipeline. Automatic systems are not always available, however. API *MPMS,* chapter 8, section 1, "Manual Sampling of Petroleum and Petroleum Products," describes the conditions under which you can obtain an acceptable manual sample from an oil or product tank: if (1) the material you're sampling contains a heavy component, like sediment and water, which clearly separates from the oil or product; (2) the tank contains either a weir or a swing suction at the bottom of the tank to prevent the shipment of the heavy component; and (3) you take the sample so that none of the heavy component is included. The standard also details the procedures used to obtain that sample. These procedures can be used for a variety of vessels, including rail cars, pipelines, marine vessels, trucks, and lease tanks.

TABLE 4.1
Summary of Sampling Procedures and Applicability

Application	Type of Container	Procedure
Liquids of more than 16 lb and not more than 26 lb Rvp[a]	Storage tanks, ship and barge tanks, tank cars, tank trucks	Precooled bottle sampling
Liquids of more than 16 lb and not more than 26 lb Rvp	Storage tanks with taps	Cooler tap sampling
Liquids of more than 2 lb and not more than 16 lb Rvp	Storage tanks, ship and barge tanks, tank cars, tank trucks	Bottle sampling; thief sampling
Liquids of 16 lb Rvp or less	Storage tanks with taps	Tap sampling
Liquids of 16 lb Rvp or less	Pipes or lines	In-line sampling
Liquids of 2 lb Rvp or less	Storage tanks, ships, barges	Bottle sampling
Liquids of 2 lb Rvp or less	Free or open-discharge streams	Dipper sampling
Liquids of 2 lb Rvp or less	Drums, barrels, cans	Tube sampling
Bottom or thief sampling of liquids of 2 lb Rvp or less	Tank cars, storage tanks	Thief sampling
Liquids and semiliquids of 2 lb Rvp or less	Free or open-discharge steams; open tanks or kettles with open heads, tank cars, tank trucks, drums	Dipper sampling
Crude petroleum	Storage tanks, ship and barge tanks, tank cars, tank trucks, pipelines	Automatic sampling; thief sampling; bottle sampling; tap sampling
Industrial aromatic hydrocarbons	Storage tanks, ship and barge tanks	Bottle sampling
Waxes, solid bitumens, other soft solids	Barrels, cases, bags, cakes	Boring sampling
Petroleum coke; lumpy solids	Freight cars, conveyors, bags, barrels, boxes	Grab sampling
Greases, soft waxes, asphalts	Kettles, drums, cans, tubes	Grease sampling
Asphaltic materials	Storage tanks, tank cars, lines, packages	—
Emulsified asphalts	Storage tanks, tank cars, lines, packages	—

Source API, *MPMS*, chap. 8, section 1, table 1.

[a]Rvp refers to Reid vapor pressure—a measure of the oil or product volatility. According to McGraw-Hill's *Dictionary of Scientific and Technical Terms*, it is "a measure in a test bomb of the vapor pressure in pounds pressure of a sample of gasoline at 100°F (37.8°C)." The higher the Rvp, the more volatile the oil or product.

RECOMMENDED PROCEDURES FOR MANUAL SAMPLING OF TANKS

You may use any of several recommended procedures to obtain a manual sample. The most commonly used include the *all-level sample,* the *running sample,* and the *composite spot sample.* The method you use depends on company policy, the equipment available, the liquid being sampled, how skilled you are at a particular procedure, and the type of storage tank being sampled. Table 4.1 summarizes the sampling procedures and their applications.

In addition to taking one of the foregoing types of samples, you will need a bottom sample to determine the height of the S&W.

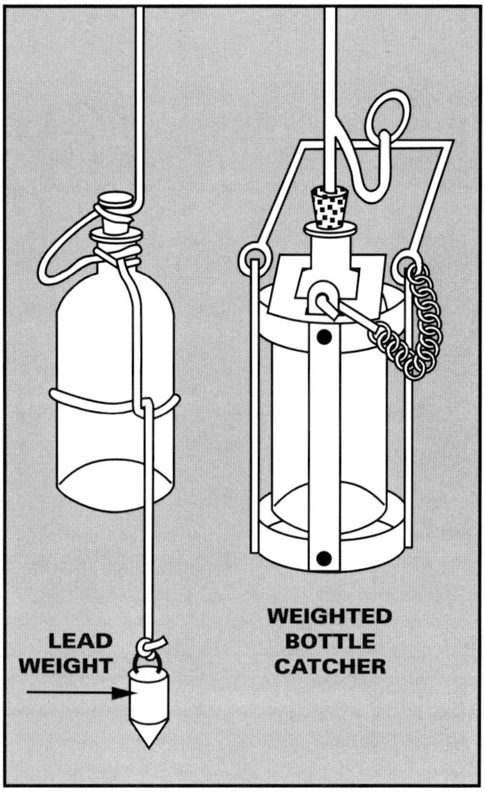

Figure 4.1. Weighted and stoppered bottles

All-Level Sample

In the all-level sample, you start with a stoppered and weighted bottle or beaker attached to a cord (fig. 4.1). The cord is attached in such a way that you can jerk it and remove the stopper from the bottle while the bottle is submerged in the oil or the product. After you stopper the bottle and attach the cord, you lower it through the hatch to a point as near as possible to the outlet, or draw-off, level. Then you jerk the cord to remove the stopper and raise the sampler at a uniform speed to the surface. A uniform speed assures that liquid replaces vapor in the bottle the entire time it is moving to the surface of the tank. The bottle should be three-quarters filled when it emerges from the liquid; if it is, the bottle was probably filling the entire time it traveled from the outlet level to the surface. If, however, the bottle is full, you can't be sure when it filled; thus you may not have obtained any of the material from some part of the tank and you don't have an acceptable sample. Pour it back and repeat the procedure but withdraw the bottle at a faster rate.

An all-level sample can be difficult to obtain. Removing the stopper near the outlet is not easy. If you do not insert the stopper into the bottle tightly enough, vapor trapped inside the bottle can force the stopper out prematurely—particularly in warm liquids. If the stopper is inserted too tightly, it is very difficult to remove. For light products or deep tanks, you may need to use a restricted opening to avoid filling the bottle.

Figure 4.2. Weighted and unstoppered bottle

Running Sample

The running sample differs from the all-level sample in that you start with a weighted, *unstoppered* bottle or beaker tied to a cord (fig. 4.2). You lower the container to near the outlet, or draw-off, level and then pull it back to the surface in one continuous movement. You move the container through the liquid at a uniform speed so that liquid replaces vapor in the bottle the entire time the bottle travels through the liquid. Just as in the all-level sample, the bottle should be approximately three-quarters filled when it surfaces. If the bottle is full when it surfaces, pour the sample back and repeat the procedure but withdrawing at a faster rate. Again, for light products or deep tanks, you may need to use a restricted opening to avoid filling the bottle.

In both the all-level and the running sample procedures, you use weighted bottles or metal beakers rather than a thief or other type of container. One advantage to using a bottle or beaker is that you can stopper the container at the surface without transferring the liquid to another container for shipment to the lab. Keeping the number of transfers to a minimum is important because you want to avoid losing the light-end hydrocarbons and S&W so that you obtain a more representative sample.

Both these procedures are difficult to perform accurately because it is difficult to raise a sample container through a liquid at a uniform speed; therefore, it is difficult to obtain a representative sample. Thus the composite spot sample is frequently used because it is easier to perform than either the running or the all-level sample.

Composite Spot Sample

The composite spot sample is a blend of several individual samples, called spot or core samples, each of which is taken at a different level in the tank. To obtain a spot sample, you use an instrument called a *thief,* or core sampler (fig. 4.3). The thief (the same instrument used to determine settled S&W levels, discussed in chapter 2) is an open-ended tube with a bottom closure. You open this bottom closure prior to lowering the thief into the tank, then lower the thief to a specified level, trip the bottom closure shut, trap the liquid, and raise the thief to the surface. You repeat this action at different levels in the tank and then combine the different samples to form one composite sample from the tank.

API *MPMS,* chapter 8, section 1, recommends that, on tanks larger than 1,000 barrels with more than 15 feet of oil, you take equal-volume samples from the midpoint of the upper, middle, and lower, or outlet, levels of the oil (fig. 4.4).

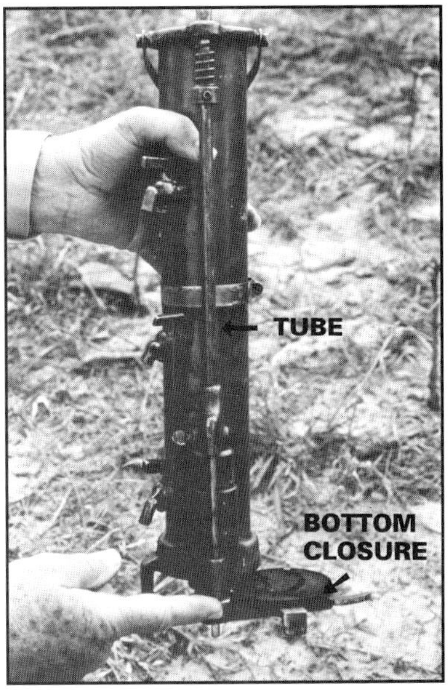

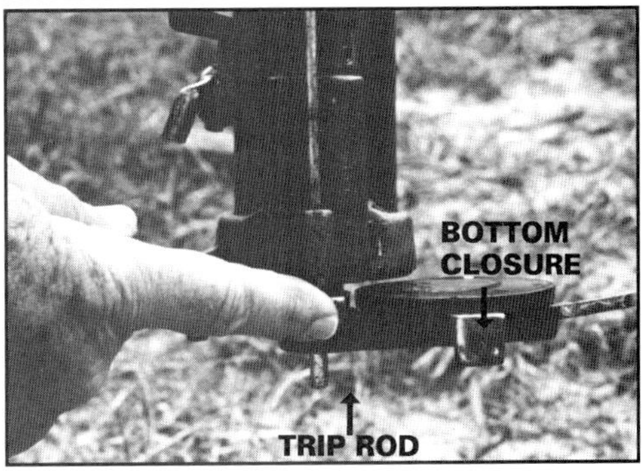

Figure 4.3. Thief, or core sampler

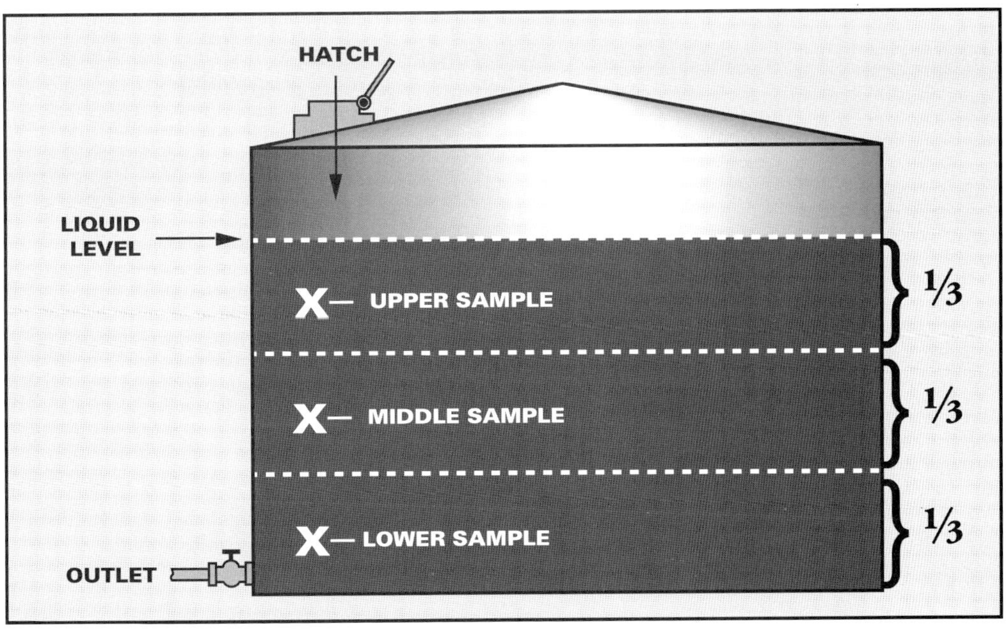

Figure 4.4. Sampling depths for composite sample

Combine the three samples to form one composite sample. On tanks holding more than 1,000 barrels but with an oil level of 10 to 15 feet, equal-volume samples should be taken from the midpoints of the upper and lower sections of the oil and combined to form one composite. If the tank contains 10 feet or less of oil, take one sample from the middle of the oil. Regardless of the number of samples taken, always take them in this order: (1) upper, (2) middle, and (3) bottom, so that you don't contaminate the remainder of the liquid.

For a tank of uniform cross section, such as an upright cylindrical tank, you blend the composite from equal parts of each spot sample. For a horizontal cylindrical tank, such as rail cars or trucks, you blend the samples in proportion to the volume of liquid sampled. Table 4.2 shows the amounts needed from each spot sample to obtain a representative blend from a horizontal cylindrical tank.

When you blend the composite, make sure you follow all safety precautions: wear gloves if necessary; minimize the number of transfers of liquid from container to container; avoid splashing the sample; avoid breathing vapors from the sample; and, if you cannot deliver the samples immediately, store them in a clean, sheltered, well-ventilated area.

TABLE 4.2
Sampling Instructions for Horizontal Cylindrical Tanks

Liquid Depth (% of diameter above bottom)	Sampling Level (% of diameter above bottom)			Composite Sample (proportion of 10 parts)		
	Upper	Middle	Lower	Upper	Middle	Lower
100	80	50	20	3	4	3
90	75	50	20	3	4	3
80	70	50	20	2	5	3
70		50	20		6	4
60		50	20		5	5
50		40	20		4	6
40			20			10
30			15			10
20			10			10
10			5			10

Source: API, *MPMS*, chap. 8, section 1.

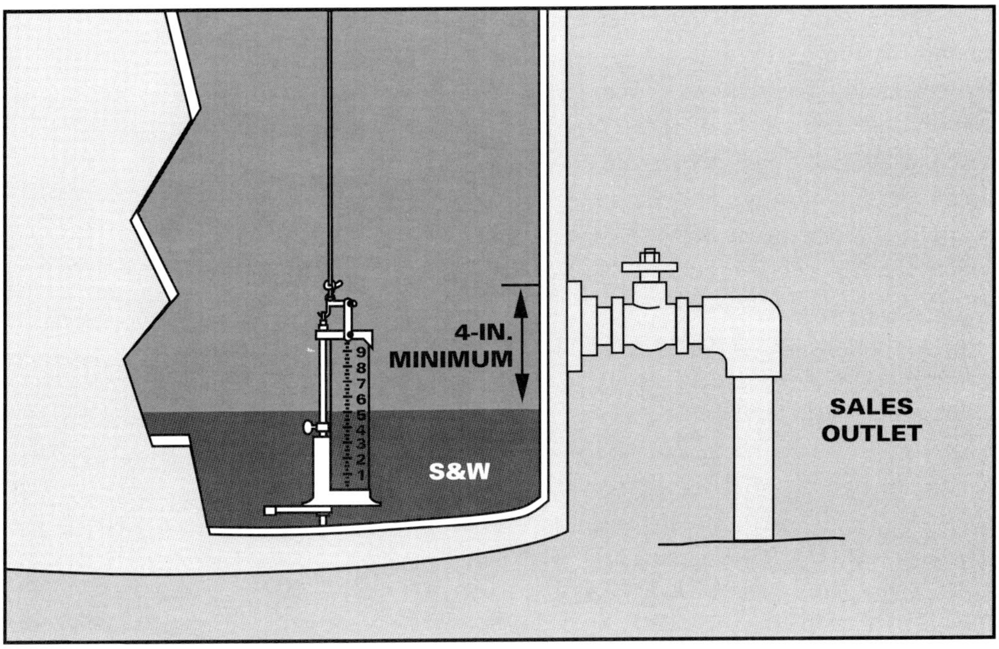

Figure 4.5. Required clearance below sales outlet

Bottom Sample

In addition to one of the foregoing methods, you should also take a *bottom sample* to determine the height of the settled S&W and to make sure that at least 4 inches of sellable hydrocarbons are below the outlet, or draw-off, level (fig. 4.5). Maintaining this 4-inch clearance helps prevent the transfer of any settled S&W.

To obtain this sample, you slowly lower the thief to the bottom of the tank. The sediment is usually heavy and viscous and may be on top of, below, or intermixed with any water present. If you lower the thief too quickly, you may disturb the S&W and alter its height. Once the thief is on bottom, allow enough time for the S&W to equalize and reach its natural level inside the thief. Then raise the thief slightly and set it down just hard enough to trip the rod and close the valve. After closing the thief, raise it to the surface.

When you retrieve the thief, it will contain not only S&W but also some sellable liquid. Slowly pour the contents of the thief over a piece of plate glass until water globules appear in the thief liquid or the liquid becomes discolored, indicating the presence of S&W. Stop pouring when you see the S&W. Return the thief to an upright position and measure the distance between the end of the trip rod and the level of S&W in the thief. This is the depth of the S&W layered on the tank bottom (fig. 4.6). Check the height of the draw-off level. The height of the S&W should be at least 4 inches below this draw-off level. If the S&W height is within 4 inches of the draw-off level, note the actual S&W height and contact your supervisor for further instructions.

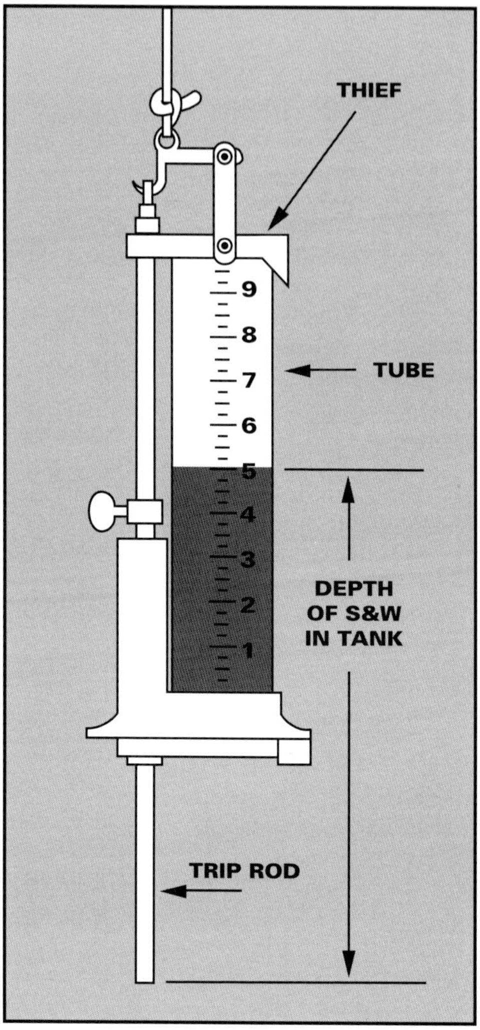

Figure 4.6. Depth of S&W is measured from the end of the trip rod to the top of S&W in tube.

TAP SAMPLING

Some tanks have taps installed on them so that you can withdraw a sample without opening the hatch. Although taps are sometimes found on crude oil tanks, they are more frequently seen on tanks containing volatile products like gasoline. Using taps and special procedures to sample volatile products helps minimize the loss of light-end hydrocarbons.

Permanent Taps

Some tanks have taps installed in the upper, middle, and lower portions of the tank (fig. 4.7). To obtain a sample from this type of tank, you should use a clean, dry, glass bottle and a clean delivery tube. The tube should be long enough to reach to the bottom of the sample container to avoid splashing the sample and losing some of the light ends (fig. 4.8). It should be free of any oil or product so that the sample will not be contaminated.

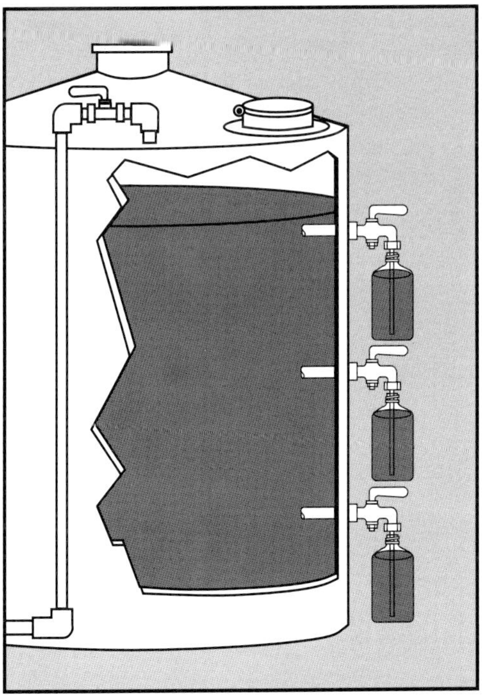

Figure 4.7. Permanent taps installed in the upper, middle, and lower portions of tank

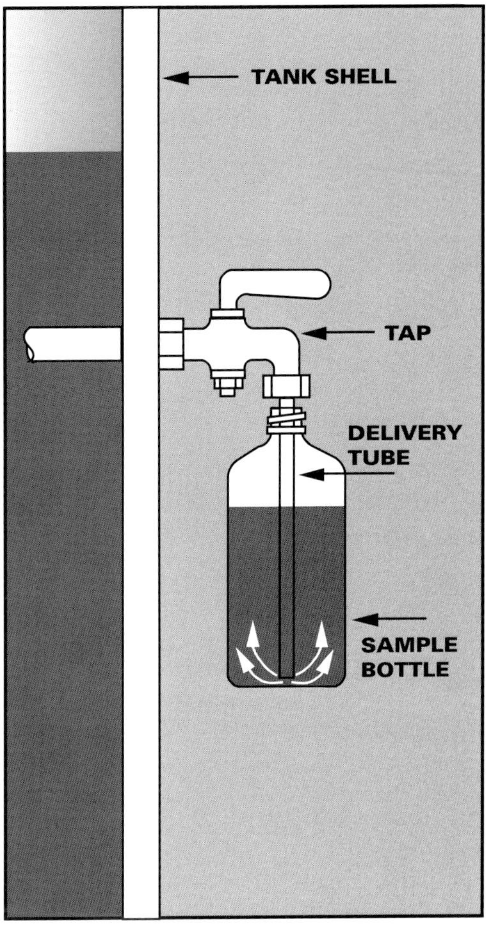

Figure 4.8. Assembly for tap sampling

Before withdrawing any sample, flush the tap to remove any product remaining in the tap line from the previous sample. Make sure you flush the liquid into a container in which it can be safely stored until properly disposed of. Using the delivery tube, fill the sample bottle. Immediately after withdrawing the sample, cap and label the container and ship it to the lab for testing. After shipping the sample, be sure you thoroughly clean the delivery tube before storing it.

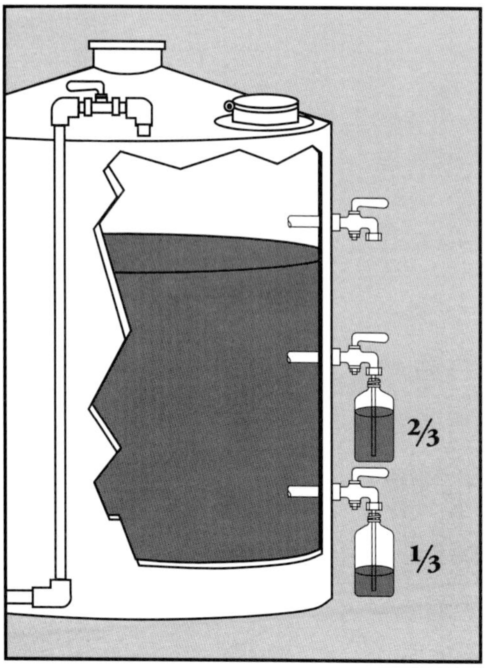

Figure 4.9. Sample amounts when level is closer to upper tap

If the tank is full, you should withdraw equal amounts of sample from each tap (fig. 4.7). If the tank is not full, you take different amounts of sample from each tap, depending on the liquid level inside the tank: if the level is closer to the upper than to the middle sample tap (fig. 4.9), take ⅔ of the sample from the middle sample tap, and ⅓ from the lower one; if the level is closer to the middle sample tap (fig. 4.10), take half of the sample from the middle tap, and half from the lower one; if the level is below the middle sample tap, take all the sample from the lower tap.

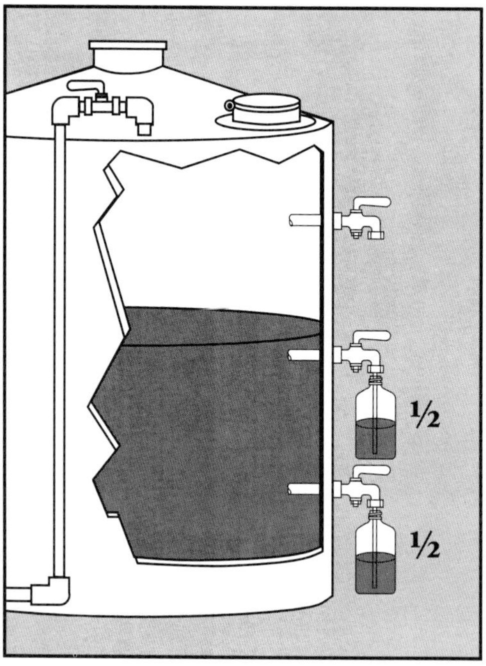

Figure 4.10. Sample amounts when level is closer to middle tap

Floating Tap Systems

Some product tanks are equipped with a floating tap system inside the tank connected to permanent taps installed outside the tank. As shown in figure 4.11, the three tap inlets are located along a floating boom or suction line. Because this boom rises and falls with the liquid level, the taps are always properly positioned in the center of the upper, middle, and lower thirds of the liquid height. With this type system, you take equal amounts of sample from each tap.

To obtain a sample from this system, flush the tap to remove any product remaining from the previous sample—just as on tanks with permament level taps. Also use a clean delivery tube to fill the container without splashing to avoid losing the light ends. And again, immediately cap and label the container, ship it to the lab for testing, and clean the delivery tube.

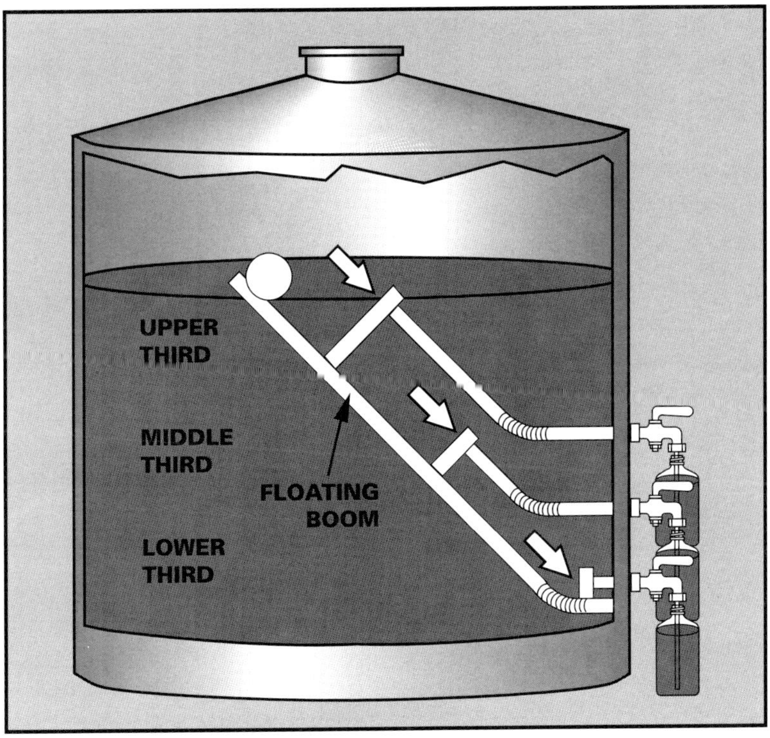

Figure 4.11. Floating tap system

MANUAL SAMPLING OF PIPELINES

Automatic sampling is preferred on pipelines carrying semiliquids and liquids of 26 pounds Reid vapor pressure or less. But if the automatic sampler fails, you may need to sample the liquid stream manually. Most pipelines have a manually operated sample probe that allows you to withdraw a sample from the midpoint of the stream. The probe should be located wherever in the line the flowing stream is adequately mixed—preferably on a vertical run of pipe.

Most pipes run along or parallel to the ground. In order to have a vertical run, the pipe must be curved at a 90° angle with a vertical run of pipe going up, followed by another 90° elbow and a straight run of pipe parallel to the ground, followed by a third 90° elbow and a vertical run back to ground level (fig. 4.12). Each time the fluid stream passes an elbow, it becomes more turbulent. Withdrawing the sample on the vertical run after the third elbow helps mix the stream and assists in obtaining a representative sample of the material passing at that instant. In addition to being placed where adequate mixing occurs, the probe should also be in a horizontal position so that none of the sample drains back into the main stream. Immediately after taking the sample from the probe, cap the bottle, label it, and ship it to a lab for testing.

The probe can be located in a horizontal run of pipe, providing the flowing velocity is high enough to provide adequate turbulent mixing of the liquid stream. If the flowing velocity is not adequate, a device for mixing the flow should be installed upstream of the sampling tap to reduce stratification. Devices used include a series of baffles in the pipe, or an orifice or perforated plate in the pipe. In addition, the pipe size may be reduced, or any combination of the foregoing may be used. See API *MPMS*, chapter 8, section 2, for specific recommendations.

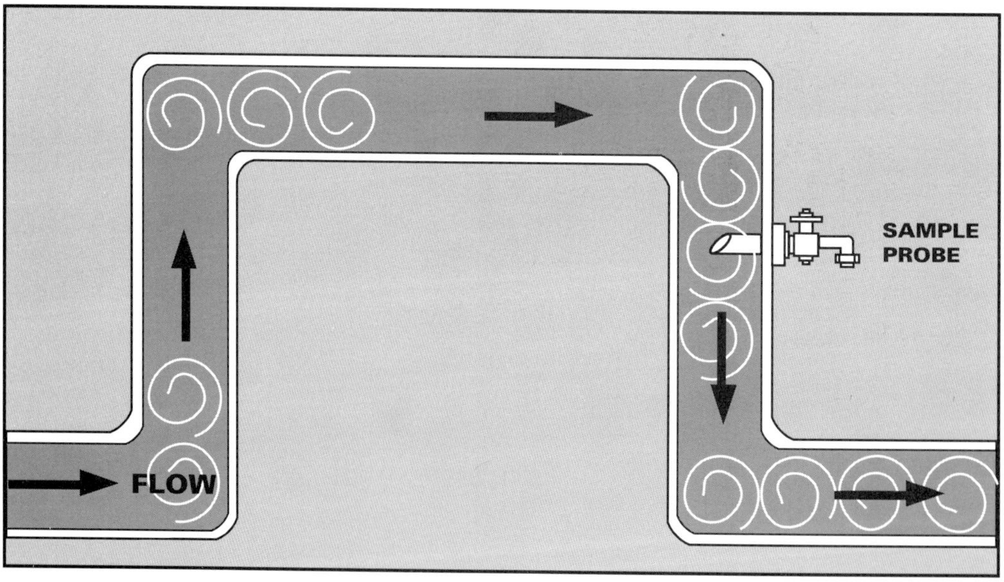

Figure 4.12. Manual probe located on vertical run of pipe

HANDLING THE SAMPLE

You should be very careful when handling any sample to make sure that it represents the general character and condition of the material. Make sure your hands are clean. Wear clean gloves only when absolutely necessary—in cold weather, when handling materials at high temperature, or for safety—to reduce the chance of introducing lint into the sample. Select wiping cloths that do not introduce lint into the sample. Many petroleum vapors are toxic and flammable, so avoid breathing the vapors and take care that you do not ignite them.

When sampling volatile products, make sure all containers have been filled with some of the product to be sampled and allowed to drain before drawing the sample. Flushing the container saturates the air inside the bottle with vapors from the product—making the environment inside the bottle similar to that in the tank—and helps you obtain a representative sample. If you transfer the sample from one container to another, make sure you have rim-to-rim contact between the two bottles until all the contents have been transferred so that no unsaturated air will be entrained in the transfer.

You should protect all volatile samples of oil or oil products from evaporation. Make any transfers immediately after withdrawing the sample. Immediately cap and keep the container closed. Be sure the container is strong enough to hold the sample during shipment. At the lab, all samples should be cooled before containers are opened.

If you are sampling liquids sensitive to light, such as gasoline containing tetraethyl lead, use brown glass bottles if available; you may use clear glass bottles, but wrap or cover them immediately. If possible, use covered cardboard cartons to house sample bottles immediately after you collect the samples.

When sampling highly refined products, you can protect them from moisture and dust by placing paper, plastic, or metal foil over the stopper and top of the container.

No matter what liquid you are sampling, never completely fill a sample container. Take into consideration the temperature of the liquid at the time you are withdrawing the sample and the probable maximum temperature to which the container may be subjected. Remember, heat causes hydrocarbon liquids to expand.

If you are using glass stoppers in your sample container and you must ship the samples to a lab, you will need to prevent the loss of liquids and vapors during shipment. Cover the stoppers of glass bottles with special plastic caps that have been swelled in water, wiped dry, and placed over the tops of the stoppered bottles and allowed to shrink into place. If the samples are to be shipped or transported over public roads or highways, they must be in containers approved by the Department of Transportation (DOT).

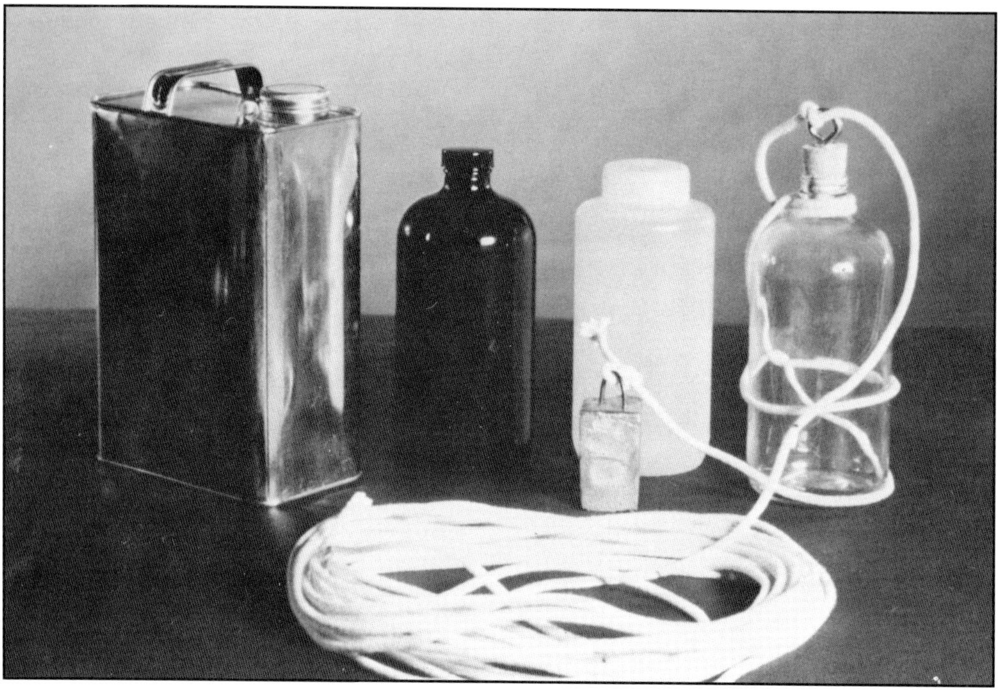

Figure 4.13. Various sample containers

MANUAL SAMPLING EQUIPMENT

Not only are there various methods of sampling, but there are also many different types of sampling equipment. The type you use depends on the sampling being done and the liquid being sampled.

Sample Containers

If the procedure and the type of liquid requires a bottle or a beaker, you can use one of several different types and sizes (fig. 4.13). If you use clear glass bottles, you can easily check the container to be sure it is clean and to see whether the sample contains free water or sediment. Brown glass bottles, on the other hand, may be difficult to see through, but they help protect certain liquids from light.

Before using a glass container, rinse it with Stoddard solvent or another naphtha of similar volatility. Then wash the container with a strong soap solution, rinse it thoroughly with tap water, and rinse it again with distilled water. If you are field-sampling crude oil, you do not need to wash the container with the soap and water. If you are sampling refined products, you may use a container rinsed with some of the product to be sampled or a sterilized container—it depends on the product sampled. You should dry the bottles either by passing a current of clean, warm air through the containers or by placing them in a dust-free cabinet at 40°C (104°F) or higher. By using hot air to dry the container, you make sure it is free of lint and dust.

You can use metal containers to store the sample if the seams are soldered on the exterior surface with a flux of rosin in a suitable solvent. Such a flux is easily removed with gasoline. Other types of fluxes may contaminate the sample and cause the lab tests to be incorrect.

You can also use special plastics for gas oil, diesel oil, fuel oil, and lubricating oil, but *do not* use them for products such as gasoline, aviation jet fuel, or kerosene, unless testing indicates there is no problem with solubility of the container or loss of light ends from the liquid. Plastic bottles have an advantage in that they do not shatter like glass or corrode like metal. You usually discard them after one use, so you do not have to worry about cleaning them.

If you use a thief to sample the liquid, wipe or rinse it clean with solvent and allow it to air dry before taking the sample.

Container Stoppers

Each container you use requires some type of stopper or cap. You can use various types, including cork, glass, or metal screw caps; however, *never* use rubber stoppers. Certain types of rubber are affected by hydrocarbons. The rubber swells or becomes soft and will not fit properly.

If you use a cork, check to make sure it is of good quality and that it is clean and free of loose bits of cork.

If you use a glass stopper, be sure it fits properly.

If you use screw caps, they should provide a vapor-tight seal. They should also be protected by a disk faced with tin, aluminum foil, or other material that will not affect the oil or product.

Cords or Lines

You should use cords—preferably cotton ones to prevent static buildup—or chains to lower and raise the sampler. The cord or chain should be marked at intervals of at least 1 foot in order to position the sampler at the proper level. Do not use gauge tape to raise or lower the sampler; the tape can become worn or kinked and will distort future measures.

CONCLUSION

Obtaining a representative sample is not impossible—it is just very important and needs to be done with care. Remember, the sample you take is used to test for specific physical and chemical characteristics and to determine price. It must be representative of the entire amount in storage, or all tests made on it will be worthless. To ensure the quality of the sample, follow the recommended practices described in this chapter and in API *MPMS*, chapter 8, section 1, "Manual Sampling of Petroleum and Petroleum Products"; use the proper equipment for the job; follow all safety procedures and company policy; and, if you have questions, get help from your supervisor.

CHAPTER 5

Automatic Sampling of Petroleum and Petroleum Products

INTRODUCTION

Millions of barrels of petroleum and petroleum products are produced and transported every day. Each fluid must be sampled in order to determine price and to test for such characteristics as the octane rating (gasolines), the sulfur content (diesel fuels), gravity (crudes), and S&W content (oil or products). The best way to obtain a representative sample of oil or product is to use an automatic sampling system.

Automatic sampling involves a constant, systematic withdrawal of small amounts from a liquid stream flowing through a pipe. If the automatic sampling system is set up correctly and maintained properly, it will give you a more representative sample of the liquid than you would get if you manually sampled the liquid. The automatic system eliminates much of the risk of human error, can be used virtually anywhere (since it is installed on pipelines), provides a constant withdrawal from any liquid stream, and reduces the personnel time required to obtain a sample. In short, it is highly efficient and accurate.

AUTOMATIC SAMPLING SYSTEM

An acceptable automatic sampling system consists of four basic parts: conditioning equipment upstream of the sampling location; devices, including a probe and an extractor, to extract fluid physically from a flowing stream; a sample receiver to collect and hold the sample grabs taken from the stream; and a controller that regulates the amount of sample taken (fig. 5.1).

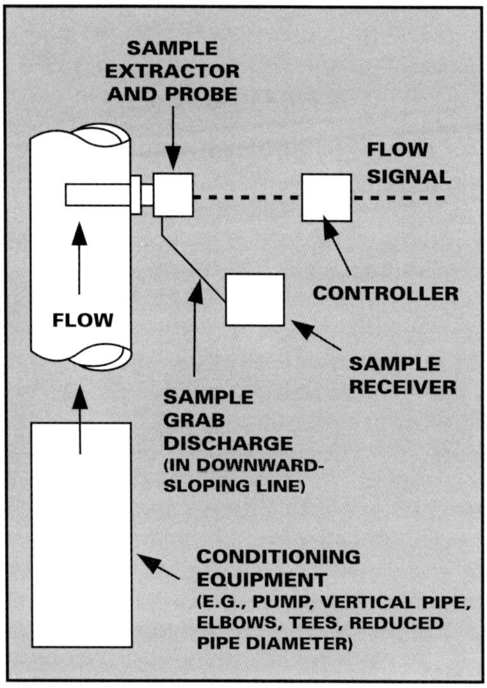

Figure 5.1. Main components of an automatic sampling system

Conditioning Equipment

If the sample taken from the liquid stream is to be representative, the stream must be mixed properly (sometimes referred to as conditioned) before the sample is taken. Basically, a sample is adequately mixed if all components of the liquid stream are distributed evenly in the stream at the sample point. Adequate mixing depends on two factors—the flow velocity and the density of the liquid. Deciding what equipment to use to obtain the required mixing for a specific liquid is the responsibility of the design engineers. They may specify using power mixers, static mixers, pressure-reducing valves, metering manifolds, lengths of reduced-diameter piping, or different piping elements or configurations (valves, elbows, tees, expansion loops). But no matter which equipment is specified for a particular job, keep in mind that its function is to distribute all components evenly in the stream.

A centrifugal pump installed upstream of the sample probe is one commonly used power mixer. The pump mixes the liquid in the pipe, but it can cause a centrifuging effect, which concentrates denser components near the sides of the line (fig. 5.2).

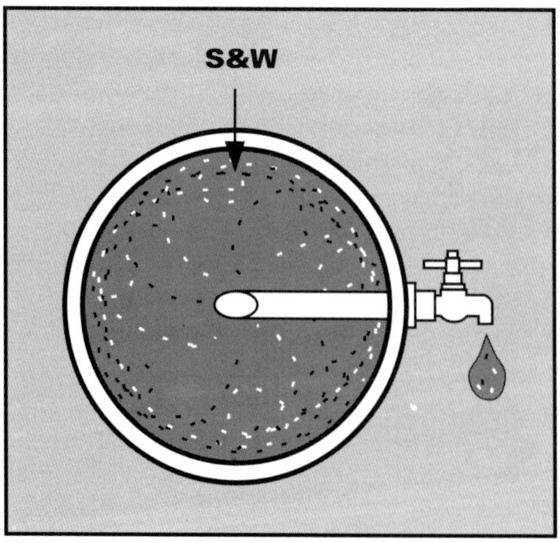

Figure 5.2. Centrifuging effect

Another way to get adequate mixing is to install the sample probe in a length of reduced-diameter pipe. Since the liquid is forced into a smaller area, flow velocity increases through the reduced section. As flow velocity increases, the flow becomes turbulent. Turbulence helps to mix the stream. One drawback to reducing pipe size is that this method also creates a permanent pressure drop that wastes horsepower and costs money.

A third method frequently used to condition the liquid stream is to provide a vertical loop in the line (fig. 4.12). As the liquid flows past each elbow in the loop, turbulence increases and mixes the stream. The sample probe should be located just before the fourth elbow to take advantage of the turbulence created by the other three. It shouldn't be located downstream of the last elbow because some settling of heavy components can occur in the horizontal section of the line.

All mixing methods have drawbacks; therefore, verify your sample results with sample proving. In this procedure you inject a known quantity of water ahead of the sampling system, take samples, analyze them for sediment and water content, and compare those results with the known water injection. This procedure is described in API *MPMS*, chapter 8, section 2.

Probe and Extractor

The probe is a small-diameter pipe that extends into the center of the liquid stream. The extractor is a mechanical device attached to the probe that grabs a small amount of liquid as it flows through the probe. Figure 5.3 illustrates how these two pieces work in combination. One type of probe has a hole extending through the probe from top to bottom for the liquid to flow through. When actuated by a signal from the controller, the sample extractor closes this hole and grabs or captures whatever liquid is in the probe at that instant. This small sample, about 1½ cubic centimetres, is discharged through a tube to the sample receiver.

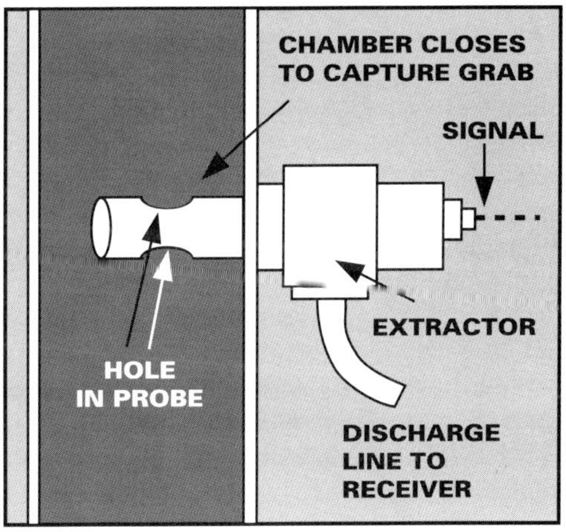

Figure 5.3. Probe and extractor

A vertical line with adequate upstream mixing is the preferred location for the probe and the extractor (e.g., on a vertical loop or on a vertical run of pipe downstream of a pump) because heavier components tend to settle more in horizontal lines. Take, for example, a liquid with S&W. The heavier S&W tends to settle. With the sample probe placed near the center of the line (fig. 5.4) and with denser components concentrated near the bottom of the pipe, the sample grab will contain less than the average amount of these components and will not be representative.

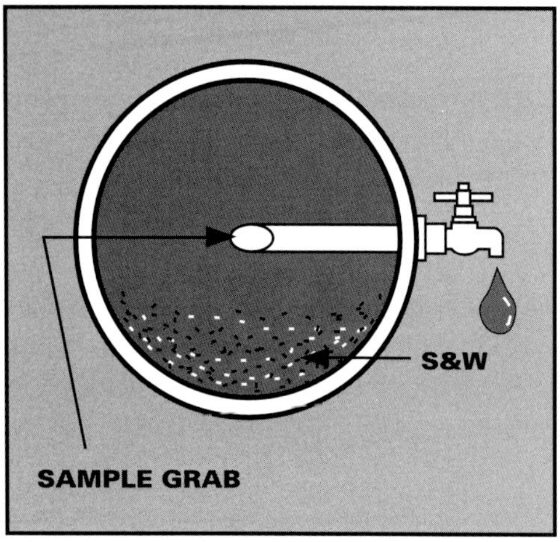

Figure 5.4. Nonrepresentative sample grab

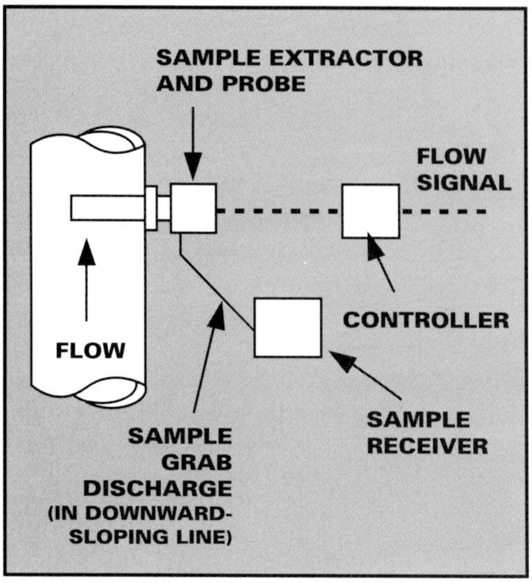

Figure 5.5. Automatic sampling—in-line

The setup and location of the probe and extractor determine what type of automatic system you have. They can be used in combination and mounted directly in the pipeline (fig. 5.5) to form an *in-line system*. Or, they can be separated, as in the *sample-loop system* (fig. 5.6). In the sample-loop system, the probe is mounted directly in the pipe but the extractor is not connected to it. The extractor is mounted on a pipe loop connected to the main pipeline. Because the extractor is located on a smaller pipe and not on the main pipeline, a sample-loop system samples a sample.

Two types of sample-loop configurations are illustrated in figures 5.6 and 5.7. In figure 5.6, a bypass line comes out of the main pipeline at the probe. The sample extractor is mounted on the bypass line just downstream of a pump that keeps flow velocity in the sample loop high enough to duplicate flow conditions in the mainline stream. A controller and a receiver complete the system.

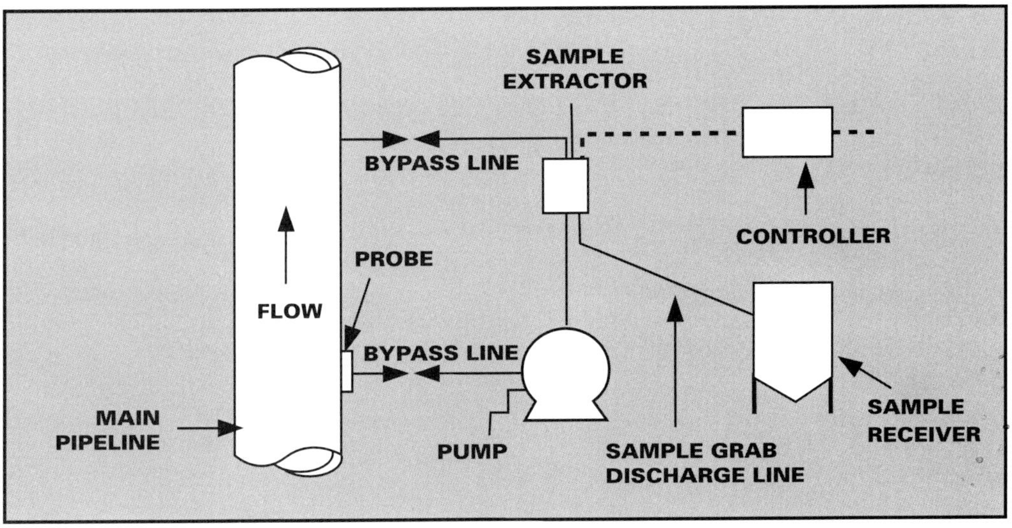

Figure 5.6. Automatic sampling—sample-loop system

Figure 5.7 illustrates another sample-loop system. This setup utilizes an existing pump in the pipeline. The loop starts at the probe on the discharge side of the pump and reenters the pipeline on the suction side of the pump. A controller and a receiver complete this setup.

No matter which sample-loop system you use, make sure the probe is installed where fluid in the pipeline is well mixed. If you are using a pump, remember to verify your sample with sample proving (see explanation of sample proving in "Conditioning Equipment" in this chapter). Pumps can force some of the heavier components of the stream to the outside edges of the pipe, thereby missing the probe.

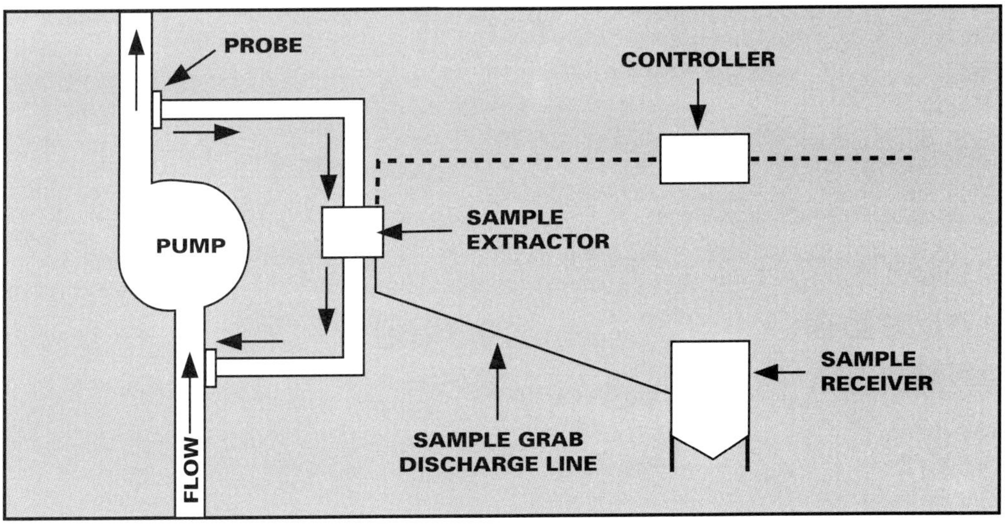

Figure 5.7. Automatic sampling—sample-loop system

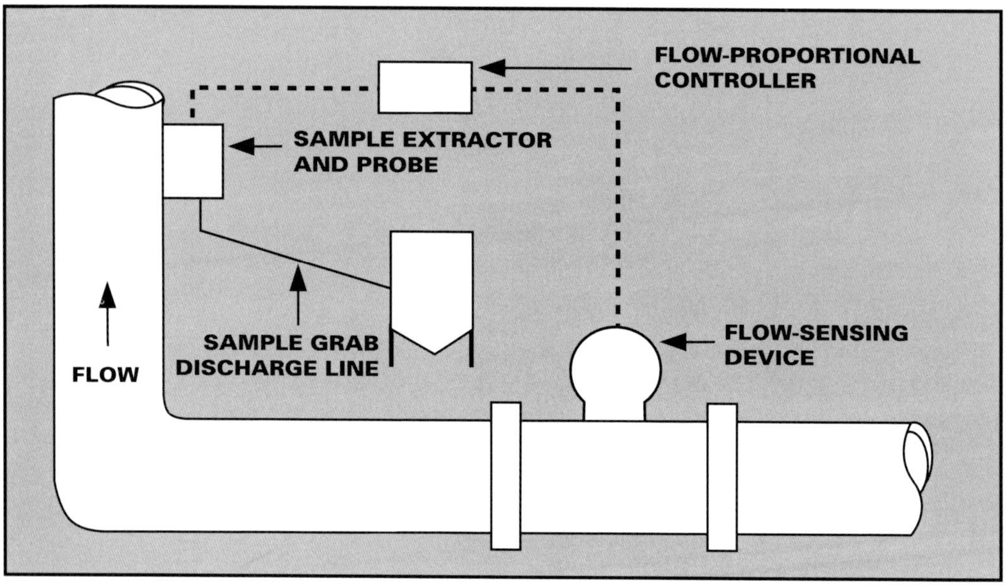

Figure 5.8. Flow-proportional sampling

Sample Controller

The controller operates the extractor. It sends a signal at regular time or volume intervals to the extractor to grab a sample. If it operates the extractor at volume intervals, the sampling is said to be *flow-proportional;* if it operates the extractor at time intervals, the sampling is said to be *time-proportional.*

If the sampling is flow-proportional, the controller is paced by a flow-sensing device in the pipeline (fig. 5.8). The flow-sensing device, through the controller, operates the extractor each time a measured volume flows past the sampling point. How frequently it signals the extractor to grab a sample depends on the sampling requirements for a given situation. For example, in pipeline operations, where S&W content tends to be fairly uniform, one grab per 320 linear feet of pipeline volume may be adequate. In marine cargo sampling, where a large part of S&W is concentrated in the line at the beginning of each discharge, one grab may be taken per 80 linear feet of pipeline volume.

How often the sample is taken depends on a lot of factors. Table 5.1 gives an idea of how sample frequency is determined. This table is just an example of the calculations needed to determine sample frequency in a marine operation. Calculations like these are necessary for each situation.

TABLE 5.1
Example Calculation for Determining Sample Frequency in Marine Operations (customary units)

A. GIVEN

Flow rate, bbl/h	30,000
Sample period, h/parcel	34
Parcel size, bbl/parcel	1,020,000
Line diameter (D), in.	36
Maximum sampler frequency (R_{max}), grabs/h	900
Grab size, ml	1.5
Lineal ft of line volume/grab (LFV)	80

B. CALCULATE

1. Minimum sample frequency, bbl/grab

$$\text{Frequency} = 0.00097\ D^2 \times LFV$$

where—

$0.00097\ D^2$ = volume, bbl, of 1 ft of pipe with a diameter of D in. (given as 36)

LFV = lineal ft of line volume/grab (given as 80).

Thus—

$$\text{Frequency} = 0.00097\ (36)^2\ (80)$$
$$= 100.6 \text{ bbl/grab.}$$

2. Number of grabs/parcel

Grabs = parcel size (given) ÷ minimum sample frequency (see step 1)

Thus—

Grabs = 1,020,000 ÷ 100.6
= 10,139 grabs/parcel.

3. Sample volume size, gallons (gal)/parcel

Volume size = [(grabs/parcel) (grab size)] ÷ 3,785 ml/gal

Thus—

Sample volume size = [(10,139) (1.5)] ÷ 3,785
= 4.108 gal/parcel.

4. Actual sampler operating rate, grabs/h

Operating rate = grabs/parcel (step 2) ÷ sample period (given)

Thus—

Operating rate = 10,139 ÷ 34
= 298.2 grabs/h.

C. FINAL CHECK

Since the maximum sampler frequency (given) is 900 grabs/h, the maximum sampler operating rate has not been exceeded.

Source: API, *MPMS*, chap. 8, section 2.

Time-proportional controllers operate the extractor at preselected times (fig. 5.9). Time-proportional sampling is used in lines in which the flow rate remains relatively constant.

Whether a system is flow-proportional or time-proportional, it needs a failure alarm. Without this alarm a large amount of liquid might flow through the pipeline without any samples being taken. Or, samples might be taken at irregular intervals, resulting in their not being representative. In addition to using an electronic alarm system, you should also visually check the sampling system in case the alarm system fails.

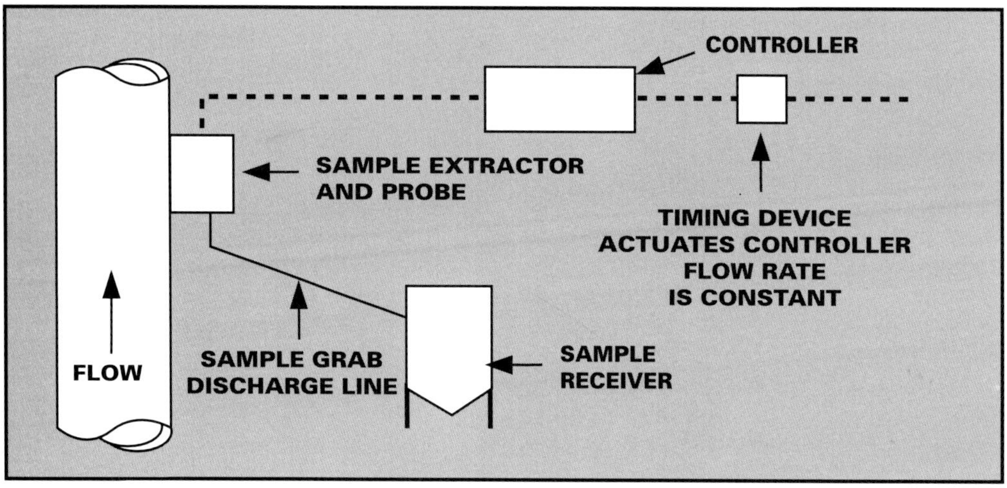

Figure 5.9. Time-proportional controller

Sampling Receivers

The receiver (fig. 5.10) is the container that holds the captured sample grabs. It should be sized to match its intended use and the operating conditions, such as parcel size (amount being delivered at a particular time), duration of the sampling period, sample grab size, sampling rate, and volume of sample required for the various users and their analyses. Some typical sample receiver sizes are 5 to 10 gallons for a LACT unit, 1 to 30 gallons for pipelines, 5 to 10 gallons at tanker loading locations, and 3 to 5 gallons at tanker unloading locations. The recommended maximum size for a portable receiver is 4½ gallons.

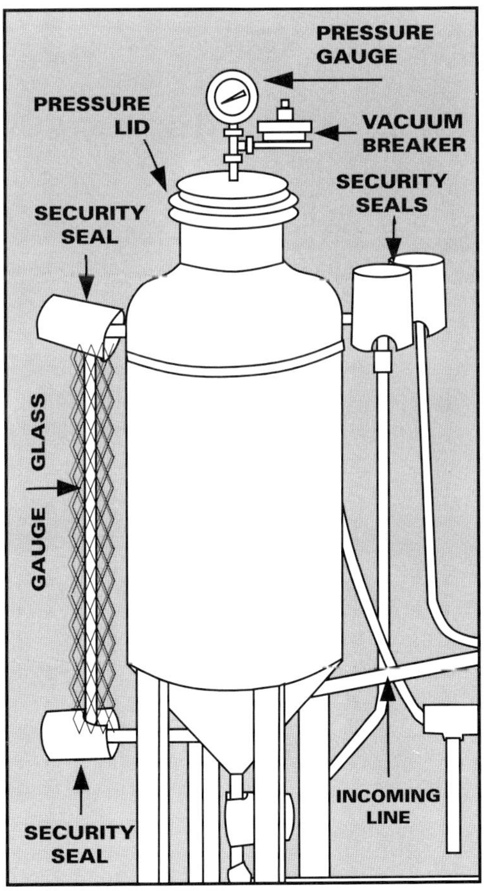

Figure 5.10. Sample receiver

Sample receivers can be permanently installed or portable, single or multiple units (fig. 5.11). Multiple units are used when different crudes or products are sampled and it is necessary to isolate them.

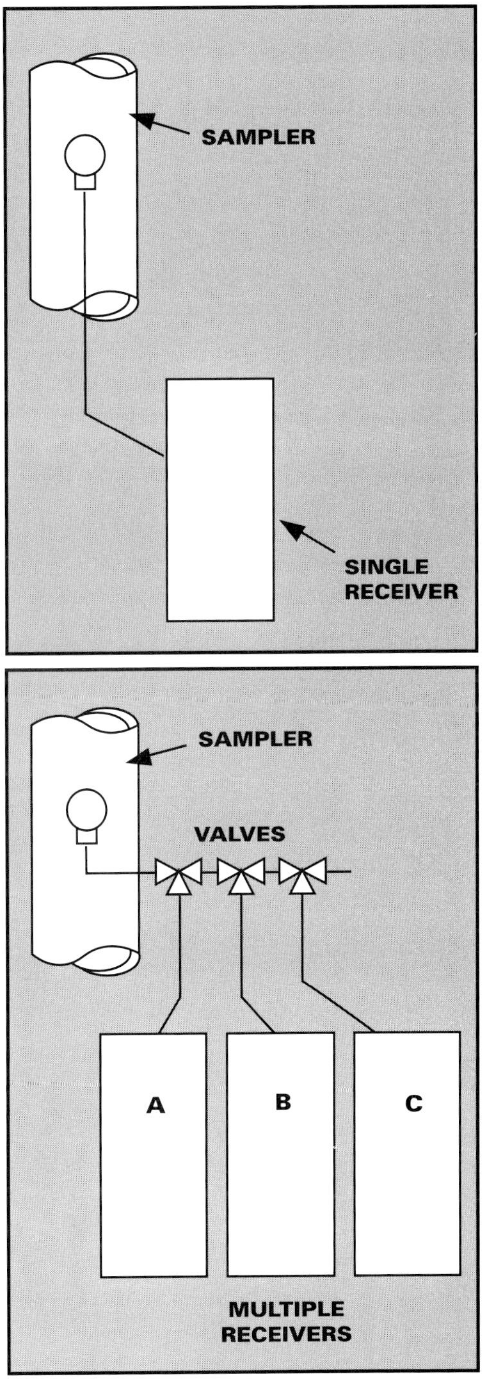

Figure 5.11. Sample receivers

Since the quality of the sample is so important, keeping it secure is also important. All receiver controls and access or withdrawal points should be sealed to prevent tampering. Piping from the sample extractor to the receiver must slope continuously downward without sags so that all the captured sample drains into the receiver. To avoid losing the light-end hydrocarbons, the receiver should be equipped with a pressure relief valve. It is normally set at about 10 pounds per square inch (psi), but it must be set at less than line pressure so that the sample can flow into the receiver.

The receiver must have a sloping bottom to aid in mixing and draining. The inside of the unit should be smooth to prevent fluids from clinging to its surface. The closure at the top of the receiver should be easy to open to facilitate cleaning and inspection. Also, no pipes to the receiver should be connected through this closure (see fig. 5.10).

A full-length gauge glass, protected to avoid breaking, should be provided to let you know that sample is being collected and to help you make sure the receiver does not overflow.

The receiver should also have a vacuum breaker so that you can drain the receiver without opening it. The breaker provides a vent to the atmosphere that you can open to break the vacuum inside the sealed receiver. The receiver should also have a pressure gauge that allows you to make sure that all openings are properly sealed and that the pressure relief valve is working properly.

Depending on the operating conditions and the material being sampled, you may need to insulate the receiver to maintain the integrity of the sample.

If you are using portable receivers, they should have carrying handles and be lightweight. In addition, they should have security seals so that all parties can be sure the sample has not been tampered with.

The receiver must be equipped with an effective power-driven mixing device to mix the sample. Hand-driven pumps or hand-driven mixing devices *will not* do the job correctly; they cannot be operated long enough to be effective.

Several types of mixer systems are available. One type of system (fig. 5.12) uses a pump to extract the liquid from the bottom of the receiver and reinject it vertically downward from the top through a perforated pipe. You can withdraw some of the sample from the valve in the discharge piping on the outside of the receiver. If this setup has suitable valves, the circulating pump can also be used to pump excess sample back to the pipeline or to a sump after you draw the portion needed for testing.

Another type uses a pump to move the liquid through spray bars (fig. 5.13). Horizontal or vertical bars will help mix the sample. With this type of system, you can draw off some of the sample from valves installed in the side of the receiver. Samples from two levels of the receiver permit comparison of results for adequate mixing.

Regardless of which mixing system is used, all piping should be arranged so that there are no low spots that can trap S&W. The piping should also permit complete drainage of the receiver, the circulating pump, and the piping itself.

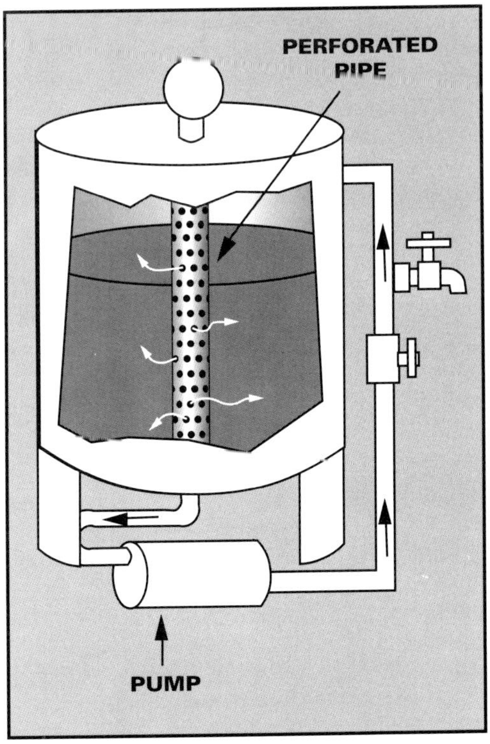

Figure 5.12. Sample-mixing device with perforated pipe

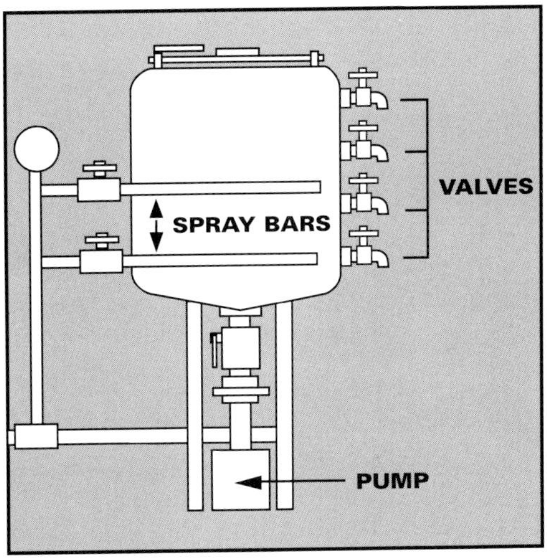

Figure 5.13. Sample-mixing device with spray bars

MIXING AND WITHDRAWING SAMPLES

Mixing times greatly affect the sample in the receiver. Different liquids require different mixing times. Be sure you know the required mixing time for the fluid being sampled. After mixing the liquid in the receiver, withdraw samples and test them before you remove any other liquid from the sample receiver. If possible, draw samples directly into the glassware in which the tests will be made; otherwise, keep transfers to a minimum. The less you handle a sample, the less chance for contamination or loss of light-end hydrocarbons.

After withdrawing the test portion from the receiver, drain and clean the receiver with a solvent. Wipe it clean with a rag, making sure that no solvent remains in it, in the pump, or in the circulating line. Any leftover solvent will contaminate the next sample.

CONCLUSION

Automatic sampling systems are preferred over manual ones because they constantly and systematically withdraw small amounts of liquid from a flowing stream. Steady, small increments sampled from a large volume where all components are well mixed will provide the best representation of the whole. If the automatic system is set up and maintained properly, it will give you what you need—an efficient, accurate, and cost-effective method for sampling oil and oil products.

CHAPTER 6

An Introduction to LACT Systems

INTRODUCTION

Manually measuring the millions of barrels of oil that are transferred from lease to pipeline each day could consume a lot of time and money. A more efficient method is to use a LACT—lease automatic custody transfer—unit. This unit measures the quantity and quality of oil (that is, it performs all the functions described in chapters 2–4) without a gauger or other personnel present. It saves the lease operator and the producing company time and money while providing a systematic means of monitoring all oil transferred from lease to carrier.

To transfer oil from lease to carrier, a LACT unit must perform three basic functions: (1) it must accurately measure the quantity of oil transferred; (2) it must sample the oil being transferred so that it can be tested to determine quality; and (3) it must monitor the S&W content to prevent the transfer of "bad oil." Bad oil contains an unacceptable amount of S&W—an extraneous commodity that is worthless and that corrodes the pipe. Keeping the amount of S&W to a minimum is necessary. What constitutes an acceptable amount is usually established by the pipeline or other carrier and is typically a little less than 1% of the total volume transferred.

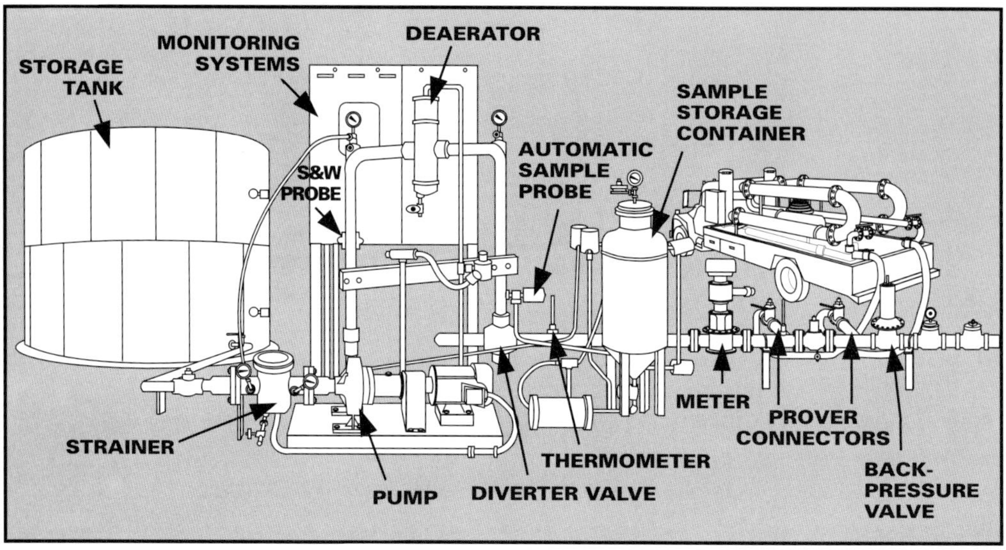

Figure 6.1. LACT unit

LACT SYSTEM REQUIREMENTS

Not all LACT systems are identical. Size, style, and configuration may differ, depending on the requirements of the situation, as long as the arrangement of parts complies with state and federal regulations. In general, however, all LACT units require storage facilities equipped with a means of controlling the flow rate from the storage unit, a centrifugal or gear pump to move the oil from lease to carrier, a probe and monitor to measure the S&W content, an automatic sampler to provide systematic withdrawal of specific volumes of liquid from the oil being transferred, a device for measuring the temperature of the oil, a diverter valve to prohibit the transfer of bad oil, a meter to record the amount of oil transferred, a fail-safe monitoring system that shuts down the unit if some part of it malfunctions, and a device that allows access for calibrating, or "proving," the meter (fig. 6.1). In addition, most LACTs are also equipped with strainers to filter debris from the oil, deaerators that allow entrained air or gas to break out of the oil and vent, and various valves, pressure gauges, thermometers, and temperature and pressure transmitters.

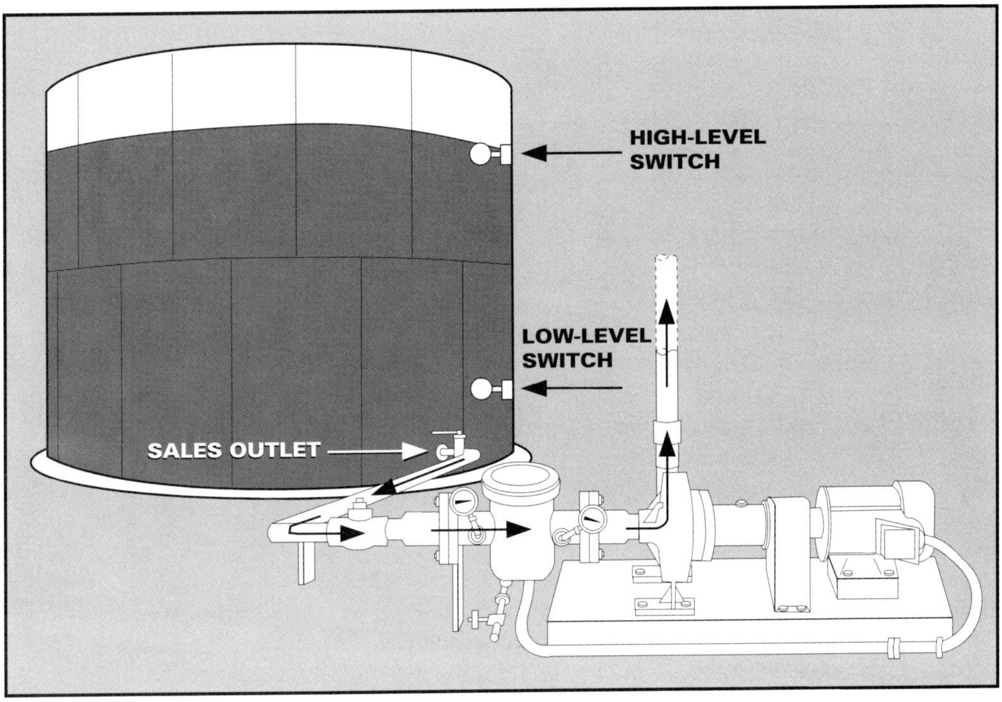

Figure 6.2. Storage or surge tank

LACT UNIT OPERATION

Since LACT units transfer oil from lease to carrier, documenting the flow of oil from storage facilities to pipeline illustrates how the unit operates.

Storage Facilities

Most oil is stored in stock tanks or surge tanks on leases. These tanks provide a place for the oil to accumulate before a custody transfer and a place to eliminate pressure surges in oil flowing from other lease equipment. Inside the tank are high- and low-level switches that control the LACT unit's operation. When the oil level inside the tank reaches the high-level switch, the pump on the LACT unit starts pulling oil from the tank. When the oil level falls to the low-level switch, the pump stops (fig. 6.2). The low-level switch is placed so that it maintains the liquid level above the sales outlet and keeps air or vapor from entering the LACT unit. Keeping air or vapor out is important because both occupy space in the line and can be measured as oil, and both can damage the LACT pump and controls.

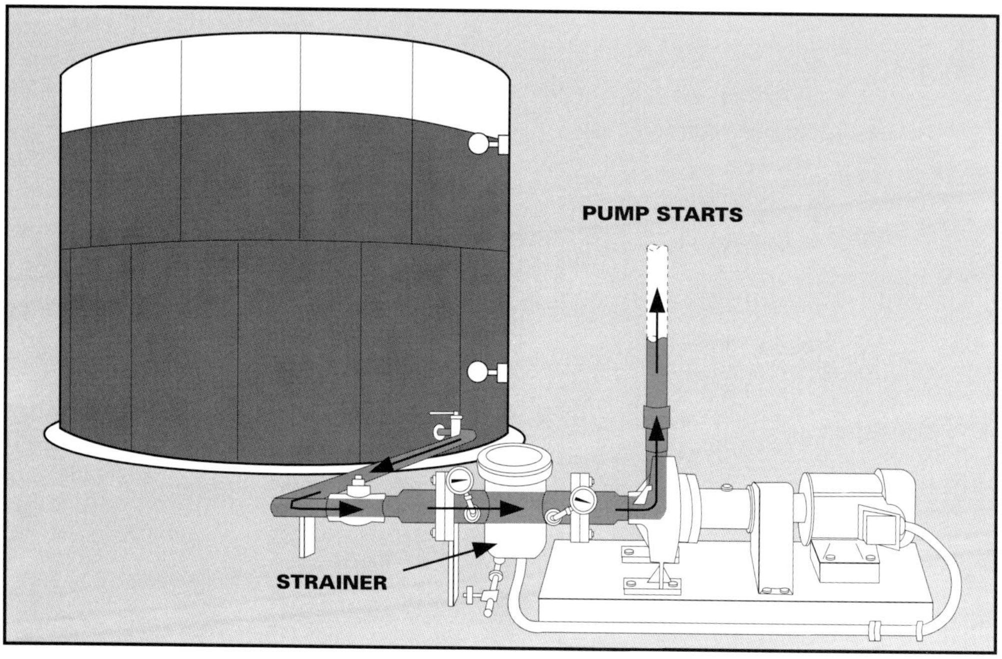

Figure 6.3. LACT unit strainer

Strainer

Most units are equipped with a strainer situated between the storage facilities and the pump (fig. 6.3). This strainer removes any large particles of sediment, sludge, or other matter from the oil that might damage the LACT or cause inaccurate measurement.

Pump

The LACT unit pump pulls oil from the storage tank and forces it through the unit at a constant pressure and flow rate. Maintaining a constant flow rate and pressure is important if measurement is to be accurate. Pumps used on LACT units are usually centrifugal or gear types, because they provide a smoother or more consistent flow than reciprocating or piston pumps (fig. 6.4) (see chap. 7 for a more detailed explanation of how pump pressure affects measurement).

Figure 6.4. Centrifugal pump on LACT unit

S&W Probe

After the oil flows past the pump, it flows past a sediment and water probe (fig. 6.5). The location of this probe (as well as the location of other LACT components) may vary, but, in each case, the arrangement of parts must comply with state and federal regulations. API *MPMS*, chapter 6, "Metering Assemblies," recommends that the S&W probe be installed in the line to the meter, since its primary function is to prevent water-contaminated oil or slugs of water from entering the pipeline or another carrier's system, and because this location usually follows the pump, which creates turbulence and helps distribute the S&W evenly in the liquid stream.

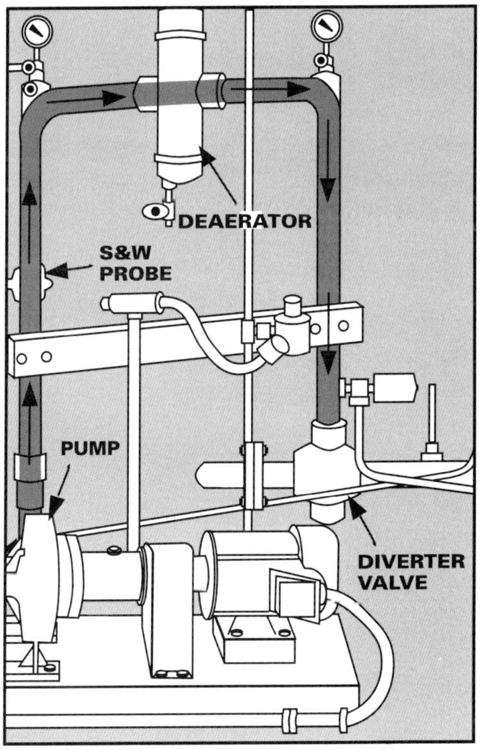

Figure 6.5. S&W probe

The S&W probe, along with the monitor attached to it, measures the capacitance, or dielectric constant, of the liquid in the line. A dielectric constant is a value of dielectricity assigned to a substance. A substance that is a good insulator has a high dielectric constant, while a poor insulator has a low one. The dielectric constant of oil is higher than that of water. The probe senses the differences in dielectric constant of the flowing stream and electronically transmits that information to the monitor. Other types of sensors are available, but the dielectric probe is the most commonly used device.

If the reading on the probe exceeds the limit set on the monitor—indicating that the water content exceeds the acceptable limit—one of two things can happen. The bad oil can be diverted to the treating system on the lease, treated to remove the S&W, and then sent through the LACT again, or the LACT unit can sound an alarm, shut down, and stop the transfer of oil before any of the liquid has been measured by the meter. If the oil is merchantable, or saleable—that is, if the reading on the probe does not exceed the limit set on the monitor, it flows past the S&W probe and monitor and on through the meter.

Deaerator

After the oil flows past the S&W probe, it usually flows through a portion of the LACT unit called the deaerator (fig. 6.6). On some units another strainer is used along with the deaerator. The deaerator slows the flow of oil so that any air or gas that is still entrained in the oil has time to break out and be vented. If the air or gas were allowed to stay in the liquid until it passed through the meter, the meter would record it as oil. In addition, the meter's rotors or impellers can be damaged if the blades hit the liquid after a large slug of air has passed through the chambers. The blades spin more rapidly as the slug of air passes through the meter. The rapidly spinning blades can then be damaged if the liquid hits the blade following the air slug.

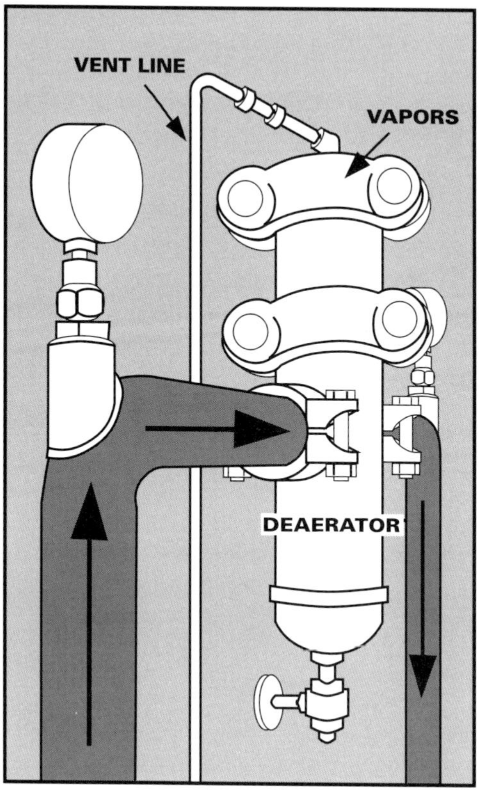

Figure 6.6. Deaerator

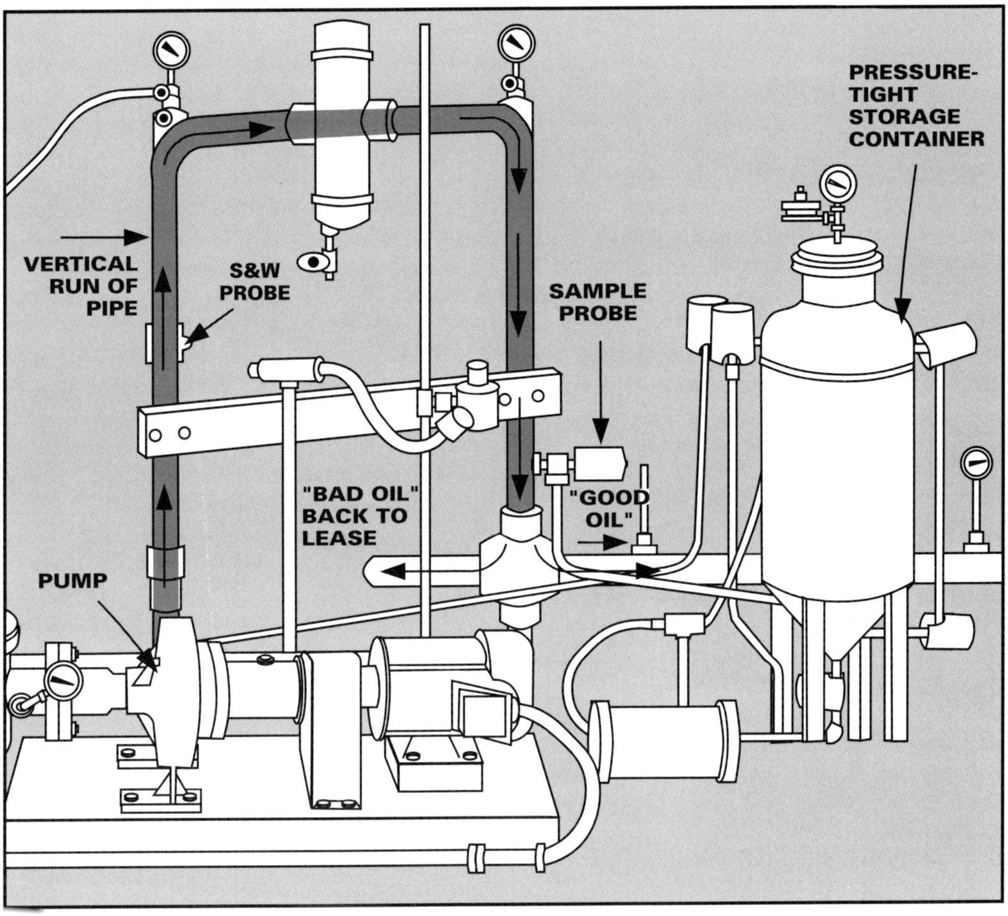

Figure 6.7. Location of sample probe in automatic sampling system

Automatic Sampling System

After the oil flows past the deaerator, it flows past a sample probe. The sample probe is located on this section of pipe so that the samples are taken after the oil passes three 90° elbows so that turbulence is high and the S&W is evenly distributed in the oil (fig. 6.7). The probe consists of a tube or plunger through which small, precise volumes of oil are taken. The probe is activated by an electronic pulse sent from the meter. The probe is set to withdraw oil at regular intervals during the transfer so that the number of samples is determined by the rate of flow passing through the meter.

The samples flow from the probe through a much smaller pipe to a nearby pressure-tight storage container (fig. 6.7). The pressure-tight seal prevents the loss of light-end hydrocarbons and keeps the samples uncontaminated until the gauger or someone else can test them. The samples must be tested for quality to verify or establish price and to determine the S&W content. Remember, any amount of S&W must be subtracted from the amount of liquid transferred. (For more detailed information on automatic sampling systems, see chapter 5.)

Meter

The meter is the single most important component of any LACT unit because it measures the volume of liquid being transferred from producer to carrier. Most LACT units use a displacement meter (fig. 6.8), in which the oil flows through the meter and turns either a rotor or lobed impellers (fig. 6.9). Each set of rotating elements forms a measuring unit of fixed volume. These measuring units break up, or displace, the liquid stream into discrete portions that fill each chamber. The rotation of these chambers, and therefore the number of portions is transmitted through connecting gears to the meter register on top of the meter to measure the amount of oil passing through the meter. (See chapter 7 for details of meter operation.)

In addition to measuring chambers, connecting gears, and register, most meters contain several other parts that are necessary for accurate measurement.

AUTOMATIC TEMPERATURE COMPENSATOR

Oil volume changes with temperature—expanding when warm and contracting when cold. The standard temperature for volume measurement is 60°F. Usually, an automatic temperature compensator (fig. 6.10) adjusts the meter register so that it converts the volume of oil transferred to an equivalent volume at 60°F instead of the flowing temperataure.

PULSE, OR FLOW, MONITOR

LACT unit meters are also equipped with a pulse, or flow, monitor. This monitor automatically shuts the LACT unit down if the meter rotation speed is too slow. Remember, the LACT unit pump must force oil through the meter within a limited range of flow rate and pressure. If the meter rotation is slower than a preset rate established for accurate measurement, the pump is not forcing the oil through the meter at the required rate. To prevent oil from being transferred when it has not been measured accurately, the flow monitor shuts down the flow through the meter and stops the custody transfer.

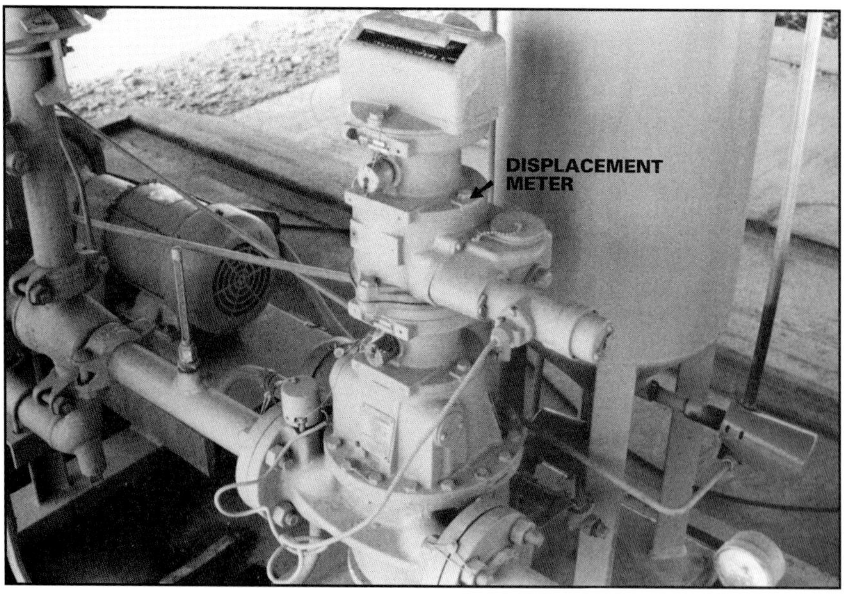

Figure 6.8. LACT unit displacement meter

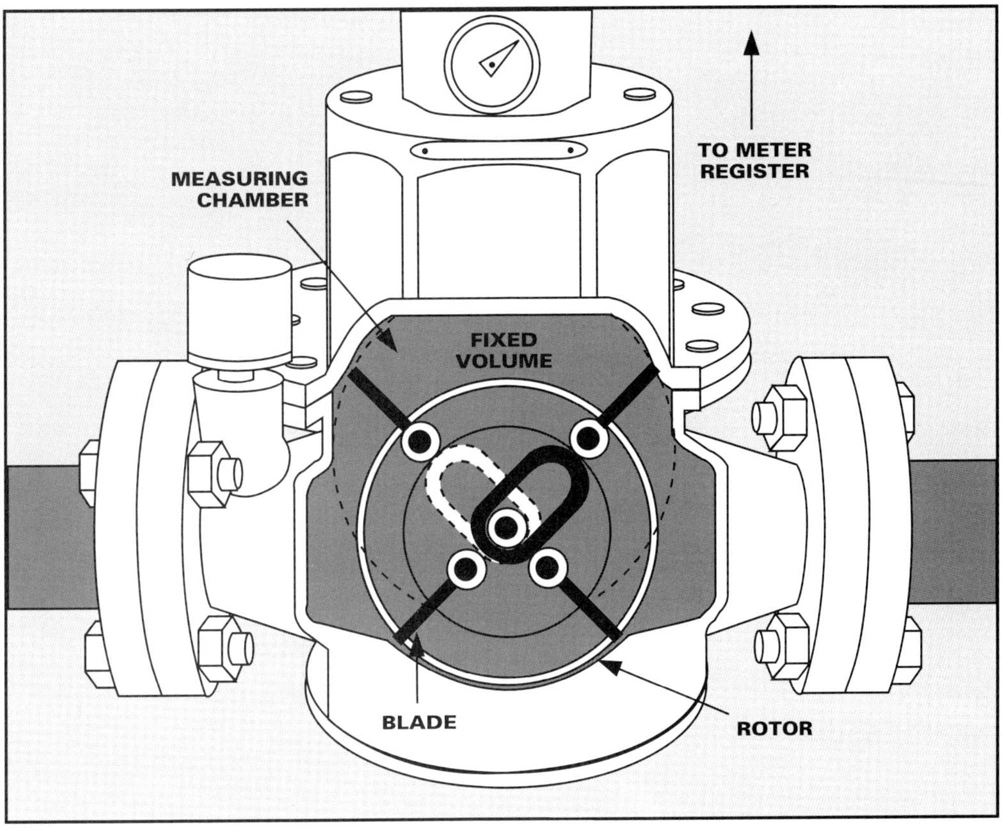

Figure 6.9. Cutaway of a displacement meter with rotor blades forming the measuring unit, or chamber

SET-STOP COUNTER

Many LACTs have a set-stop counter in the meter stack or in a meter control panel attached to the meter. This counter shuts the LACT unit down once the maximum allowable is reached. The *allowable* is the amount of oil that may legally be removed from a field over a certain period of time.

Figure 6.10. Automatic temperature compensator

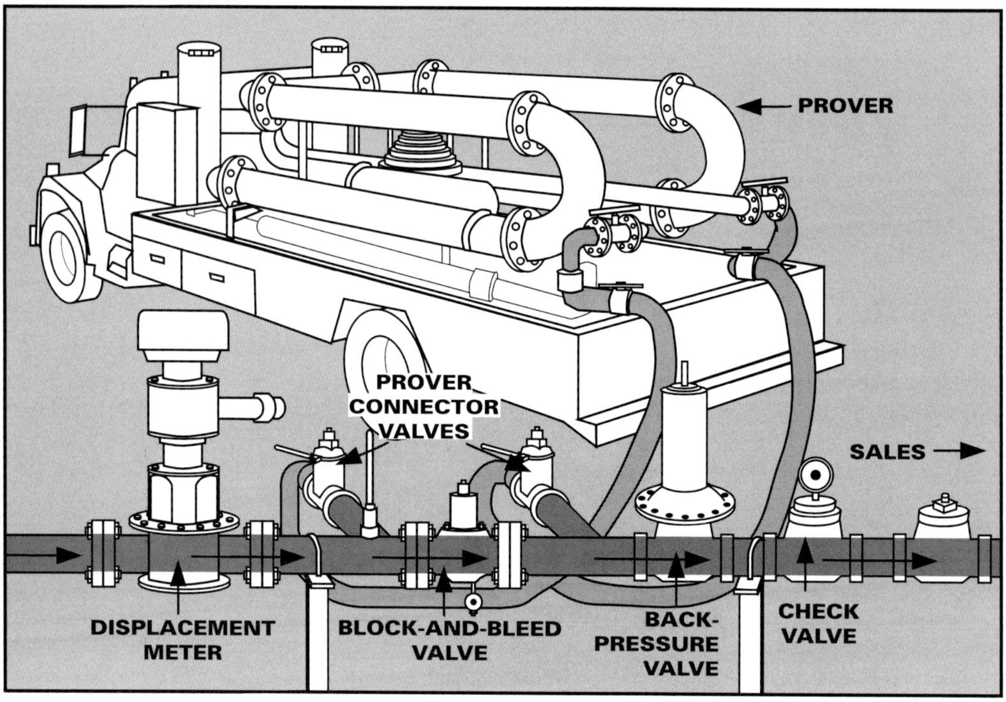

Figure 6.11. Prover connectors

Prover Connectors

After going through the meter, the oil flows past two prover connector valves and through an open block-and-bleed valve (fig. 6.11). These valves provide access for meter proving.

During proving, the oil is blocked from the sales line by the block-and-bleed valve and forced through a prover, a device of known calibrated volume. Provers are often portable, mounted on trucks or trailers, and driven to various sites to prove the meter (fig. 6.11). The liquid volume passing through the prover and the volume registered on the meter are then compared. Several mathematical calculations are made, and a meter factor is determined. Because of design and manufacturing limitations and wear and tear, all meters have a certain margin of error. The numerical meter factor is calculated to correct this error. This factor is used to correct each meter reading to reflect true volume. This factor must be determined on a regular schedule because it does not stay the same for the life of the meter. (See chapter 7 for a detailed explanation of how a prover works and how meters are proved.) Changes in flow rate, environmental changes, wear and tear, changes in liquid viscosity, and other factors all affect the margin of error and thus the meter factor.

Back-Pressure Valve

During normal operations, the oil flows through the open block-and-bleed valve past a back-pressure valve. This valve is adjusted to help maintain a constant pressure on the LACT unit while the pump is running.

Check Valve

After flowing past the back-pressure valve, the oil flows past a check valve. This valve keeps oil from flowing back into the LACT unit when it is not in operation (fig. 6.12).

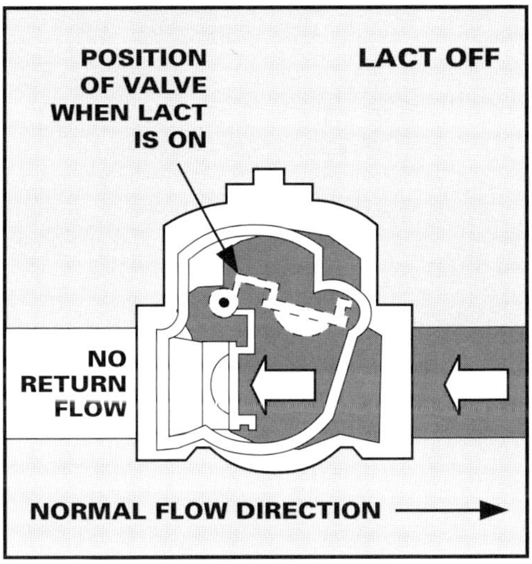

Figure 6.12. Check valve

CONCLUSION

LACT units save a great deal of time and effort for the gauger or any others working the lease. As long as the unit is maintained in operating condition and the meter regularly proved, it will accurately measure the amount of liquid transferred, provide automatic samples for testing, and monitor the S&W content, thereby preventing the sale of bad oil.

CHAPTER 7

Meters and Meter Proving

INTRODUCTION

A *meter* is a device that measures the volume of a fluid flowing through it. Many types are used in a variety of locations, from the lease to truck-loading racks, to measure oil or oil products, but the two most common types used to measure liquids are the displacement meter and the turbine meter. All meters, however, regardless of type or location, measure and register how much oil or product has been bought, sold, or transferred to another's custody. And they must measure this liquid volume accurately.

Even the best meter, no matter how well designed, installed, and maintained, does not always measure *exactly* the amount flowing through it. Each has a particular margin of error, whether the result of design limitations or of wear and tear.

Determining the meter's accuracy, or *proving* the meter, is a very important job in the oil industry. Each meter must be checked periodically, or proved, to determine its margin of error, that is, the difference between the amount registered and the true volume flowing through it. The registered volume can be corrected by manually adjusting, or calibrating, the meter, or by determining a numerical meter factor that is used in a mathematical formula to convert the meter's registered volume to volume indicated by the prover.

In this chapter the displacement meter and the turbine meter will be described as well as two of the most commonly used proving procedures—proving a meter with an open-tank prover, and proving a meter with a pipe prover.

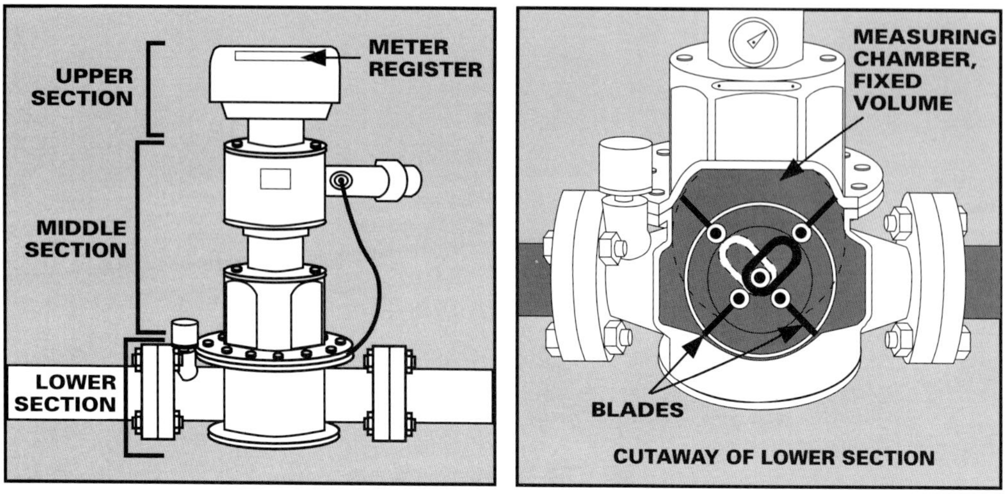

Figure 7.1. Displacement meter

DISPLACEMENT METERS

Displacement meters are one of the most commonly used meters in liquid measurement. They are one of the oldest of the mechanical devices used to measure liquids and have proven to be reliable, in spite of certain limitations. Figure 7.1 illustrates the basic parts of a common type of displacement meter. A lower section houses a rotor and blades, the middle section houses gears connecting the rotor to the register, and the upper section houses the meter register. Liquid flows through the lower section and turns the rotor and blades. As the rotor turns, the blades, or sliding vanes, and the housing form a seal. When the seal is formed, the liquid between the blades is isolated from the rest of the stream in a measuring chamber. In other words, the meter separates the flowing stream into discrete volumetric portions. The number of these portions is transmitted through the gears in the middle section to the register above, which indicates the amount of liquid flowing through the meter.

The operation of a displacement meter is similar to that of a revolving door. As people stream through a revolving door, they fill the chambers between each dividing section of the door (the equivalent of the meter's blades). The door *displaces* the stream of people into discrete groups. Each group rotates around the center of the door and exits the other side. If you had a counter connected to the door, you could count the number of people going through the door by counting the revolutions. For example, if the revolving door had 5 sections and 1 person fit into each section, then 1 revolution of the door would mean 5 people had gone through the door, 30 revolutions would mean 150 people had passed through the door, 60 revolutions would indicate 300 people, and so on.

Suppose, continuing the door analogy, that a number of people avoided the revolving door by going through the side doors instead of waiting to fill one of the sections. These people would not be accounted for by counting the revolutions of the door. The more people who slipped through the side doors, the more inaccurate counting the revolutions would be. The same principle applies to displacement meters. In order for the rotor and blades to turn, there must be a space, or clearance, between the rotating parts and the meter housing. Some liquid slips through these clearances and is not registered. This *slippage* is accounted for during regularly scheduled meter provings. Pumping the liquid through the meter at a consistent rate is important because it keeps the amount of slippage from varying. An increase in flow rate increases the amount of slippage; a decrease in flow rate decreases the amount of slippage; maintaining a steady flow rate through the meter helps maintain a consistent rate of slippage, and thus helps accurately account for all oil flowing through the meter (fig. 7.2).

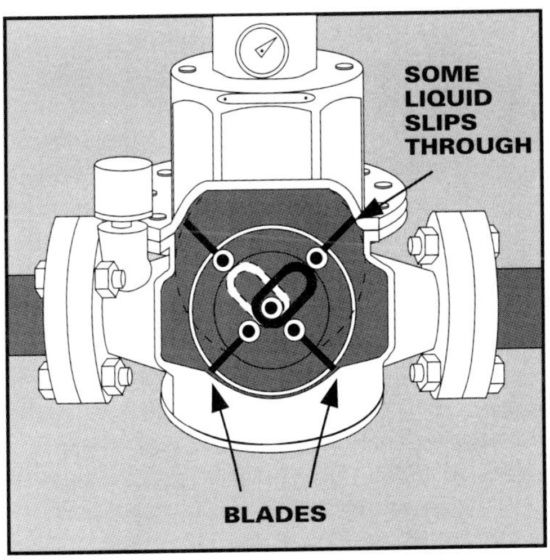

Figure 7.2. Slippage through a displacement meter

Another problem with this type of meter is wear. As the rotor turns, wear on the moving parts causes changes in the meter's performance: clearances tend to increase, and gears and other moving parts become worn and therefore less precise. The result is that more oil or product slips through the meter unregistered.

Despite these limitations, this type of meter is used throughout the industry. It is regularly proved to determine just how much oil has passed through unmeasured and to account for the discrepancy when determining true volume through the meter. Also, the displacement meter is generally found to be more accurate than are turbine meters with more viscous hydrocarbons like heavy crude oil. Finally, slippage may allow some liquid through the meter that is not registered, but it also allows the sand and the sediment in some crudes to flow though the meter without undue damage.

Another type of displacement meter (fig. 7.3) uses lobed impellers to displace the liquid stream into discrete segments that are then counted and registered. The impellers work like gears that mesh and turn to create measuring chambers. The liquid stream flowing through the meter forces a lower impeller with rounded projections to rotate. As this impeller turns, its projections fit into the concave sections of the upper impeller, forcing it to turn. As the upper impeller turns, its concave portion and the housing meet to form a measuring chamber. The number of turns of this impeller and thus the number of discrete sections of liquid flowing through the meter is transmitted through gears to the meter register.

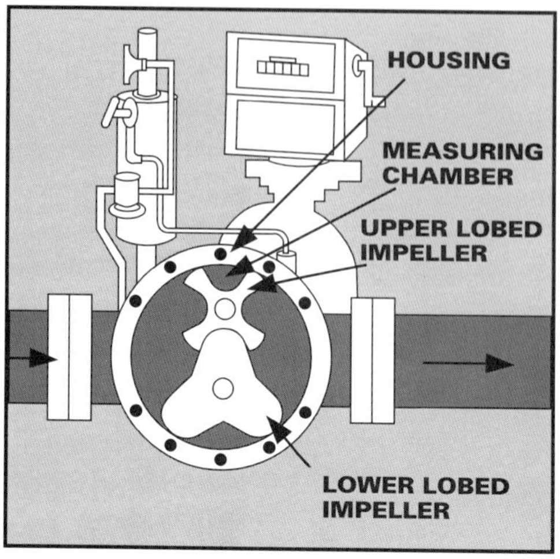

Figure 7.3. Displacement meter with lobed impeller

The lobed impeller meter also has design limitations that affect its accuracy. Like the bladed, or sliding vane, meter, it, too, requires some clearance between the impellers and the housing. This allows liquid to slip through unregistered. Also like the bladed meter, wear and tear will cause the moving parts to become worn. And, fluctuating flow rates will mean varying rates of slippage and, therefore, inaccurate measurement. This meter, too, has been used extensively despite its limitations because regularly scheduled meter provings determine its margin of error and compensate for the difference between registered volume and true volume.

TURBINE METERS

A turbine meter measures liquid flowing through it in a completely different way from the displacement meter. Instead of displacing the liquid stream into discrete parts and measuring the number of parts, the turbine meter measures the velocity of the liquid moving through a known area (meter) and infers the volume flowing through it. Because turbine meters measure the velocity of the liquid and from that measurement infer the flow rate and thus the volume, they are referred to as inferential meters. The speed or flow rate of a liquid is directly proportional to its volume—the higher the flow rate through a given area, the greater the volume. Because the turbine meter uses the speed of the liquid stream to determine volume, it is sometimes called a *velocity meter.*

Figure 7.4 illustrates the basic parts of a turbine meter, including the rotor, shaped much like a windmill, the stator (hanger or diffuser), the pickup coil, and a register or indicator. Liquid flowing past the rotor turns the rotor, just as wind will turn the blades of a windmill. Each blade of the rotor is counted and registered. The more liquid flowing through the meter, the faster the rotor turns and the higher the numbers on the register. Conversely, if the flow rate declines, less liquid flows through the meter, the rotor does not turn as often, and the register indicates less volume through the meter.

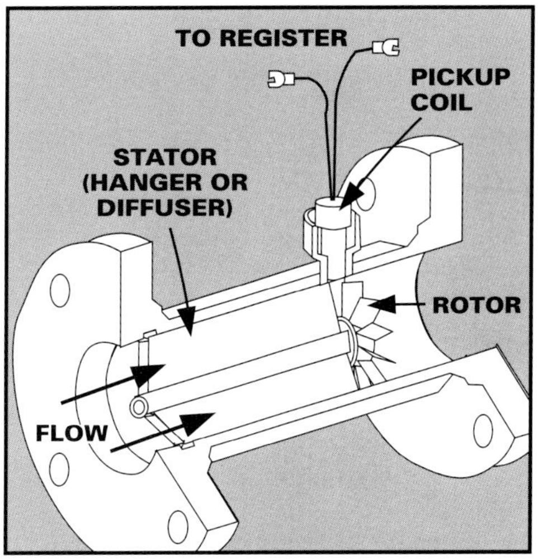

Figure 7.4. Turbine meter

Turbine meters are very accurate with the lighter hydrocarbons like gasoline, diesel fuels, and distillates. But wear and tear affects this type of meter, just as it does any other. If the rotor or blades are damaged or worn, they do not perform as well as they did when new. Thus, just as with any other type of meter, the difference between the registered volume and the true volume must be verified or proved on a regular schedule.

BASICS OF METER PROVING

Proving meters is, in theory, at least, a simple procedure: you compare a meter's registered volume with a known volume. To do this, you precisely measure a volume of liquid either before or after it is run through the meter. Once you know the difference between the meter's registered volume and the true volume, you calculate a special number—the *meter factor*—and then use that number to adjust the meter electronically or mechanically so that it registers true volume or use the meter factor to correct each meter reading mathematically. You can use the meter factor mathematically to correct with both displacement or turbine meters, but you can adjust displacement meters only mechanically.

Mechanical and electronic adjustment and mathematical correction are all acceptable ways to ensure that the meter is reflecting true volume. Deciding on which method to use usually depends on the location of the meter. Displacement meters that handle relatively small amounts of liquid, like those at truck-loading facilities, for example, are often mechanically adjusted in increments of 5⁄10,000 (.0005)%. At a truck stop loading a 2,000-gallon truck compartment, adjusting the meter in increments this small can reduce the margin of error to only 1 gallon. Since truck transactions are rounded to the nearest gallon, this error is insignificant. On the other hand, meters at locations handling large volumes of oil or product must be as close to 100% accurate as possible because even a small percentage error can mean a much larger loss when dealing with thousands of barrels of liquid. For example, if a meter is handling a run of 200,000 barrels, mechanically adjusting the meter in .0005 increments can mean an error of 100 barrels. One hundred barrels is a significant error, so meters that measure large volumes of liquid typically use the meter factor with a mathematical formula to correct the registered volume to true volume.

As stated earlier, proving a meter is in theory a simple process—determining a known volume and comparing that amount with a meter's registered volume. To accomplish this task, however, you must use fairly complicated equipment, be thoroughly trained in a precise list of procedures, and take into consideration the effects of pressure and temperature on liquid and steel volumes. As is often the case, reality is more complicated than theory.

Two commonly used provers are the open-tank prover and the pipe prover. A master meter can be used to prove another meter, but it is not the most common means. A master meter is simply a calibrated meter used instead of a prover. You flow liquid through the master meter either upstream or downstream of the meter being proved and compare the registered volumes.

Figure 7.5. Portable open-tank prover

Open-Tank Prover

A portable open-tank prover is shown in figure 7.5. There are other types, but this type is often used to prove meters at truck-loading racks because using it is similar to loading a tank truck. This prover is not under pressure: it is either open to the atmosphere or to a vapor recovery system.

To prove a meter with an open-tank prover, liquid flows through the meter and then into the prover. The amount registered on the meter is compared with the amount shown by a graduated gauge glass on the open-tank prover (fig. 7.6). The graduated glass should be read to the nearest 0.02% of the total prover volume.

Figure 7.6. Graduated glass on an open-tank prover

Pipe Prover

Pipe provers are completely different from open-tank provers. One common type is shown in figure 7.7—the bidirectional U-type. There are other pipe provers, but this type is often used to prove meters in pipelines because it is shaped like and operates much like a pipeline. The liquid stream can be diverted through it and then back to the pipeline, and because of its configuration, it is portable and fairly compact. To prove a meter with a bidirectional U-type pipe prover, you divert flow from the main pipeline into the prover—either before or after the liquid goes through the meter—then compare the volume registered on the meter with the amount that flowed though the pipe prover to determine the margin of error. Details of how this comparison is made are discussed in "Proving Meters with Pipe Provers."

Converting Volumes to Standards

With both the open-tank prover and the pipe prover, you compare the prover's known volume with the meter's indicated volume. Before you can compare these two figures, however, you must first convert both volumes to what they would be at standard temperature and pressure (usually 60°F and 0 psig). You convert these volumes to standards because oil and oil product volumes change with temperature and pressure, and all parties must agree on exactly how much liquid has flowed through the meter. For example, liquid volumes expand as temperatures rise and contract as temperatures fall—10,000 barrels of oil at 75°F occupy about 10,050 barrels at 85°F. Liquid volumes also expand as pressure decreases and contract as pressure increases—10,000 barrels of oil at 500 psig occupy about 10,030 barrels at 0 psig. The size of the prover also changes with temperature and pressure changes.

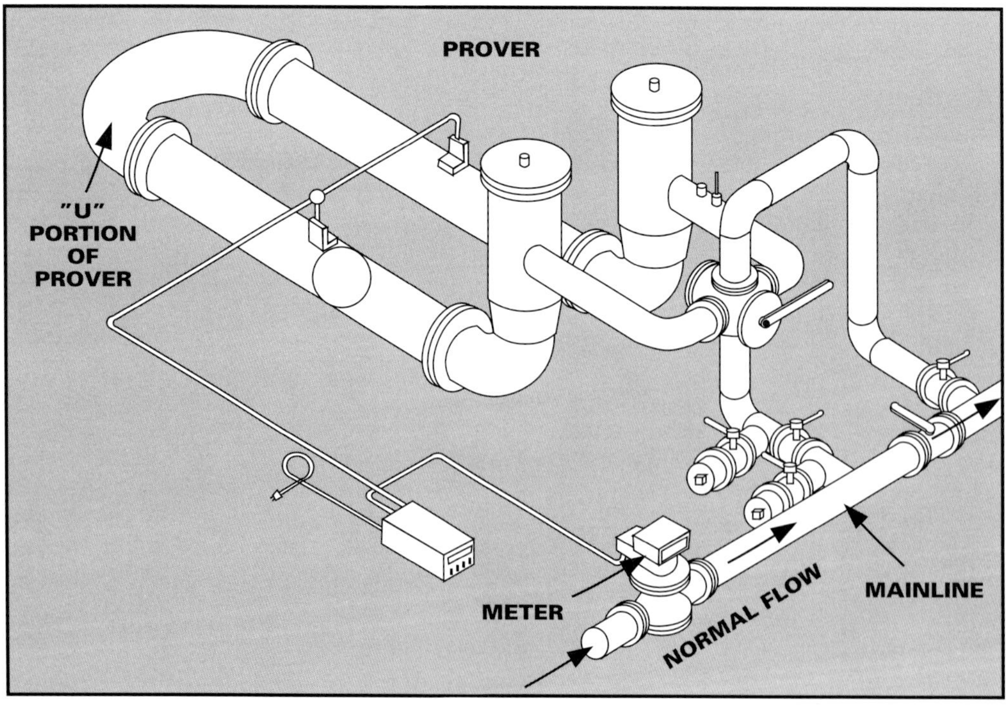

Figure 7.7. Bidirectional U-type pipe prover

Thus, regardless of what the volumes of both prover and meter are at the ambient temperature and flowing pressure, they must be converted to what they would be if they were at 60°F and 0 psig.

In order to make these conversions, you multiply the *observed volumes* at prover and meter by certain correction factors:

1. a correction for the temperature of the steel (prover only), C_{ts};
2. a correction for the pressure on the steel (prover only), C_{ps};
3. a correction for the temperature of the liquid (prover and meter), C_{tl}; and
4. a correction for the pressure on the liquid (prover and meter), C_{pl}.

API provides tables with correction factors for temperature and pressure for both liquid and steel. Table 7.1 is an example of one of these tables. To find a correction factor on this table, you locate the column for observed temperature and then move across to the next column to locate the correction factor. For example, if the observed temperature of the prover steel is 78°F, the C_{ts} value is 1.0003 because 78° falls between 73.5° and 78.8°. Once you have located the correct numbers for each factor, you multiply the observed volume by all the required factors to convert that volume to standard temperature and pressure. The sequence of multiplication and round-off procedures is standardized so that accounting values are repeatable. API specifies these procedures.

TABLE 7.1

Temperature Correction Factors for Mild Steel Having a Cubical Coefficient of Expansion of 1.86 × 10^{-5}per °F

Observed Temperature, °F	C_{ts} Value	Observed Temperature, °F	C_{ts} Value
–7.2 to –1.9	0.9988	73.5 to 78.8	1.0003
–1.8 to 3.5	0.9989	78.9 to 84.1	1.0004
3.6 to 8.9	0.9990	84.2 to 89.5	1.0005
9.0 to 14.3	0.9991	89.6 to 94.9	1.0006
14.4 to 19.6	0.9992	95.0 to 100.3	1.0007
19.7 to 25.0	0.9993	100.4 to 105.6	1.0008
25.1 to 30.4	0.9994	105.7 to 111.0	1.0009
30.5 to 35.8	0.9995	111.1 to 116.4	1.0010
35.9 to 41.1	0.9996	116.5 to 121.8	1.0011
41.2 to 46.5	0.9997	121.9 to 127.2	1.0012
46.6 to 51.9	0.9998	127.3 to 132.5	1.0013
52.0 to 57.3	0.9999	132.6 to 137.9	1.0014
57.4 to 62.6	1.0000	138.0 to 143.3	1.0015
62.7 to 68.0	1.0001	143.4 to 148.7	1.0016
68.1 to 73.4	1.0002	148.8 to 154.0	1.0017

Source: API, *MPMS,* chap. 12, section 2.

Notes: This table is suitable for application in meter proving; in prover calibration, use the formulas. For the formula used to derive the tabulated values and to calculate values, see API, *MPMS,* chap. 12, section 2, subparagraph 5.

You do not always have to determine all four correction factors for each observed volume. Sometimes the meter or the prover already compensates for one or more factors. For instance, if you are proving with an open-tank prover, you are mainly concerned with C_{ts} and C_{tl}, the temperature factors, because the liquid is under no pressure except for the atmosphere—it is already at 0 psig. And, since the meter is calibrated at atmospheric pressure, no pressure corrrection is needed for it either. The temperature factors are very important, however. Typically, there is a difference in temperature between the prover and the meter. A meter at a load rack, where piping from the storage tanks to the meter is underground and the prover is exposed to the sun, can differ as much as 2°F from the prover.

In pipe proving, you use all the correction factors unless the meter is already temperature-compensated. Some meters, regardless of their location, have a device that automatically corrects the metered volume to standard temperature. If the meter you are proving has this device, you do not have to determine the C_{tl} for the meter. These devices do require, however, that you dial in the API gravity of the liquid so that the temperature adjustment for that particular liquid is correct.

Be sure you know which correction factors you need to determine in order to convert the observed volumes at prover and meter to standard temperature and pressure. Once both volumes have been converted, you can compare the prover volume to the meter volume and calculate a meter factor. Then you either mechanically adjust the displacement meter or use the meter factor mathematically to correct the meter reading to reflect true volume.

REQUIREMENTS FOR PROVING

Regardless of the type of proving method used, you must fulfill certain requirements to ensure valid results. First, prove the meter under the same conditions under which it operates. If, for example, the flow rate differs from normal operations during proving, the results may not be accurate for the operating flow rate. The same can be said for a change in temperature, pressure, or type of liquid flowing through the line.

Second, make sure the prover has a large-enough capacity to provide *proving runs* of adequate duration. The duration of various proving runs is discussed in detail in "Proving Meters with Open-Tank Provers" and "Proving Meters with Pipe Provers."

Third, make sure that you make enough proving runs to establish repeatability; that is, prove the meter enough times to make sure your results are accurate. In order for the proving results to be accurate, the margin of error must repeat several times in succession. Different companies have different requirements regarding repeatability. Check to be sure how close each margin of error must be and how many times it must repeat for the proving to be accurate. For example, in figure 7.8, 10 runs were made before 5 were close enough to meet the company's requirements. API recommends that, with pipe provers, 5 consecutive runs be within a span of 0.0005 to determine a meter factor for a particular proving; with tank provers it recommends 3 consecutive runs within a span of 0.0005 to determine a meter factor for a particular proving.

Fourth, the prover must be calibrated with NIST's field-standard test measures and procedures. A certificate of calibration should accompany the prover.

FLOW RATE	NON-RESET TOTALIZER
900 bbl/hr.	

RUN DATA

TEMPERATURE		PRESSURE		TOTAL PULSES	RUN NO.
PROVER AVG.	METER	PROVER	METER		
72	TC	88	92	~~85,402~~	1
72	TC	88	92	~~85,372~~	2
72	TC	88	92	~~85,357~~	3
72	TC	88	92	~~85,360~~	4
72	TC	88	92	~~85,422~~	5
72	TC	88	92	85,360 ✓	6
72	TC	88	92	85,357 ✓	7
72	TC	88	92	85,359 ✓	8
72	TC	88	92	85,360 ✓	9
72	TC	88	92	85,359 ✓	10
72	TC	88	92	85,359	AVG.

Figure 7.8. Part of a proving run report

Fifth, make sure you fill out the proving report (fig. 7.9) carefully and date and sign it. The report includes the information needed to determine a meter factor and becomes a record to use in oil and oil product transactions and for meter evaluation.

Evaluating a meter's performance is very important. All companies require that the meter operate within certain limits, or *tolerance*. If the meter factors vary a great deal between provings (at the same flowing conditions), the meter is said to be out of tolerance and may be worn and in need of repair. For example, suppose you prove a meter and determine a meter factor of 0.9984 and compare that with the previous meter factor of 0.9888. For your company the difference between the two is within the specified limits. But suppose the next time you prove this meter the factor is 0.9940—quite a bit different from the previous two and out of tolerance for your company. You must find out why the meter factor was so different from the previous provings.

The first thing you do is make sure your proving job was good and that the difference is not due to your error. If the proving job was good, you should check flow rate, temperature, and gravity of the liquid for any changes. A flow rate change can affect the meter's performance and affect its meter factor. In displacement meters increased flow rate typically increases the meter factor and decreased flow rate decreases the meter factor. With a low flow rate, the amount of liquid that slips through the clearances in the meter is low; that is, the meter factor decreases. With a high flow rate, the amount of liquid slipping through the clearances increases and the meter factor goes up.

Changes in temperature can also affect the meter factor. Oil and oil products generally become less viscous, or thinner, with increases in temperature. Less-viscous liquids slip through the clearances more easily than thicker ones; therefore, the higher the temperature, the higher the meter factor.

Changes in the API gravity of the liquid also change the meter factor because, in general, a high-gravity liquid is less viscous than a low-gravity liquid and slips through clearances more easily.

If no changes in any of these conditions account for the meter's being out of tolerance, then the meter may need repair or replacement. In general, a meter in constant use wears faster than one used only occasionally and a new or overhauled meter wears faster than one that's been in use for some time.

METER PROVING REPORT FOR TANK PROVER METHOD

LOCATION	TENDER	LIQUID	°API	DATE	AMBIENT TEMP.	REPORT NO.
Anywhere	13	Unleaded Gasoline	60.3	3/18	53 °F	42

PROVER DATA

PREVIOUS REPORT NO. ______

NOMINAL VOLUME AT 60°F AND "0" psig.	SERIAL NO.	FLOW RATE	FACTOR	DATE
1000 gal/bbl	12345	15.46 bbl/hr	—	2/15

METER DATA

SERIAL NO.	METER NO.	TEMP. COMPENSATED	MANUFACTURER	SIZE	MODEL
67890	4	☐ YES ☒ NO	X	4"	Z

FLOW RATE	NON RESET COUNTER	REMARKS, REPAIRS, ADJUSTMENTS, ETC.
15.45 bbl/hr	—	Normal Flow Rate Constant

	PROVER TANK VOLUME DATA	RUN NO. 1	RUN NO. 2	RUN NO. 3	RUN NO. 4
1	DELIVERY TO TANK, gal/bbls	1000.15	1000.10	1000.45	
2	TANK TEMPERATURE (AVERAGE) °F	54.0	54.0	54.0	
3	C_{ts}	.9999	.9999	.9999	
4	C_{tl}	1.0024	1.0024	1.0024	
5	COMBINED CORRECTION FACTOR (LINE 3 × LINE 4)	1.0023	1.0023	1.0023	
6	CORRECTED PROVER VOLUME (LINE 1 × LINE 5)	1002.45	1002.40	1002.75	

	PROVED METER DATA	RUN NO. 1	RUN NO. 2	RUN NO. 3	RUN NO. 4
7	FINAL METER READING	1001.82	1000.00	1000.33	
8	INITIAL METER READING	0000.00	0000.00	0000.00	
9	INDICATED VOLUME BY METER, bbls (LINE 7 − LINE 8)	—	—	—	
10	INDICATED VOLUME BY METER, gals (LINE 7 − LINE 8) OR (42 × LINE 9)	1001.82	1000.00	1000.33	
11	TEMPERATURE AT METER, °F	54.0	54.0	54.0	
12	PRESSURE AT METER, psig	5.0	5.0	5.0	
13	C_{tl} USE 1.000 IF TEMP. COMPENSATED	1.0024	1.0024	1.0024	
14	C_{pl}	—	—	—	
15	*CCF* (LINE 13 × LINE 14)	1.0024	1.0024	1.0024	
16	CORRECTED METER VOLUME (LINE 10 × LINE 13)	1004.22	1002.40	1002.73	
17	METER FACTOR (LINE 6 ÷ LINE 16)	0.99823	1.0000	1.00001	

METER FACTOR (AVERAGE VALUE)		C_{pl} LIQUID CORRECTION FOR PRESSURE AT METERING CONDITIONS		COMPOSITE FACTOR USE FOR CONSTANT PRESSURE APPLICATION
0.9994	X	—	=	0.9994

SIGNATURE	DATE	COMPANY REPRESENTATIVE
R K Davis	3/18	Ace Oil
B L Smith	3/18	Device Distributing

(API MPMS Chapter 12.2) 30

Figure 7.9. Meter proving report for tank prover method

Figure 7.10. Leak detector valve

Sixth, check your prover to be sure it is in good condition and is operating the way it should. Be sure that no leaks occur during a proving run. One indication of a leak is the lack of repeatability of proving runs. You can check for external leaks by simply looking at all connections. For internal leaks you need internal leak detectors. Figure 7.10 shows a small valve at the base of the main valve on a prover. To check for internal leaks, open this small valve after the main one is seated, or closed. If liquid flows out of the small valve, then the main valve is leaking and needs repair before proving can continue.

It is very important that even small leaks be fixed, even though the volumes used in proving a meter are relatively small. With an open-tank prover, a leak of 1 gallon for a 1,000-gallon proving run may not seem like much, but if you ignore the loss of 1 gallon per 1,000 in a proving run, the error adds up to 1,000 gallons for every million gallons through the meter—quite a significant error. Even the leakage of 1 pint per 1,000 gallons is significant. A pint of leakage adds up to 125 gallons for every million gallons. Even 125 gallons is too much to ignore. With pipe provers, where the measuring section for one proving run can be about 5 barrels or less, leakage is even more critical. For a prover of this size, the leakage of less than 1 ounce can cost you 1 barrel per 20,000 or 100 barrels per 2 million.

The thermometer on any prover must give accurate temperature readings. Even a 1°F error with liquids that range from 30° to 65° API gravity can be significant. Depending on the gravity, the correction for 1° of temperature ranges from 4 to 6 barrels per 10,000; therefore, be sure the thermometer is installed correctly and is not damaged. You should also compare its readings with those of a calibrated thermometer. If the two readings do not agree, then repair or replace the prover thermometer.

Air or vapor flowing with the liquid through the meter can cause proving errors. Air and vapor, like the liquid, impart motion to the meter's rotor. Thus, the air or vapor is registered as though it were liquid. The air or vapor also increases slippage and thus increases the meter factor. If the air or vapor goes through the metering system in slugs, the meter probably will not repeat within tolerance, which will give you an indication of its presence. But, if the air or vapor is evenly dispersed in the liquid, it may go undetected during the proving and the meter may or may not repeat within tolerance. In any case, if air or vapor is present, it must be eliminated. The meter and the prover should measure liquid, not air or vapor. In an open-tank prover, you can actually see the air or the vapor in the liquid through the sight glass on the top neck of the prover. In a pipe prover, you cannot see the air or the vapor, but you can tell if it's there when you bleed the prover between proving runs. If air or vapor is there every time you bleed the prover, you have a problem. You should check all possible sources of the air or vapor and solve the problem. The gland or the seal on the suction side of a pump may be leaking and pulling air into the line. Other connections and seals may also allow air into the system. Whatever the source of the problem, repair it before attempting the proving.

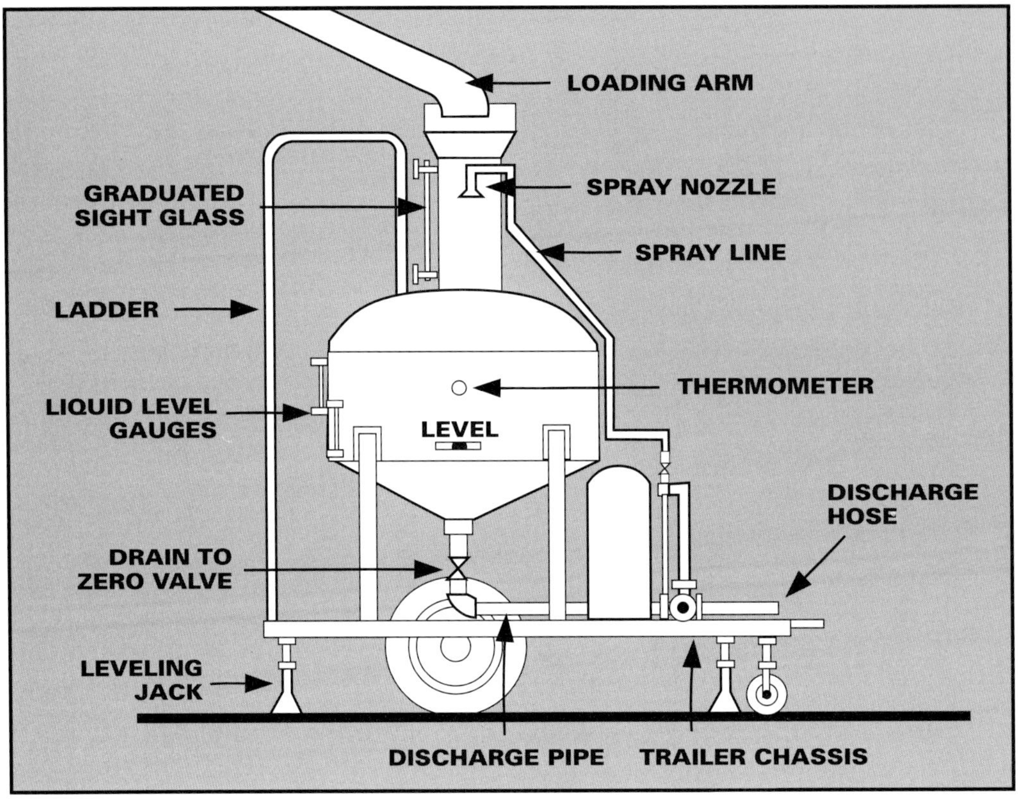

Figure 7.11. Portable open-tank prover

PROVING METERS WITH AN OPEN-TANK PROVER

As stated earlier, open-tank provers are most often used at truck-loading racks, where a large number of daily transactions are made, because using a tank prover is similar to loading a truck. Figure 7.11 illustrates the basic parts of most open-tank provers. Open-tank provers may be portable or permanently installed. There is not much difference between the two except that the portable is mounted on a truck bed or trailer and can be driven to various sites.

Both portable and permanently installed provers can be either dry-bottom or wet-bottom types. The main difference between these two types is the location of their respective gauge glasses and the technique used to drain liquid from them. Both types have liquid-level gauge glasses on the neck of the prover to indicate when the prover is filled. The wet-bottom prover has an additional gauge glass at the bottom (fig. 7.12). The techniques used to drain these provers are discussed in "Preparation for Proving."

Some open-tank provers have a top-spray device that is used when the prover is being emptied. The device saturates the air inside the prover with the vapor of the test liquid to reduce evaporation. All provers must be stabilized with a test liquid before a proving run can be made.

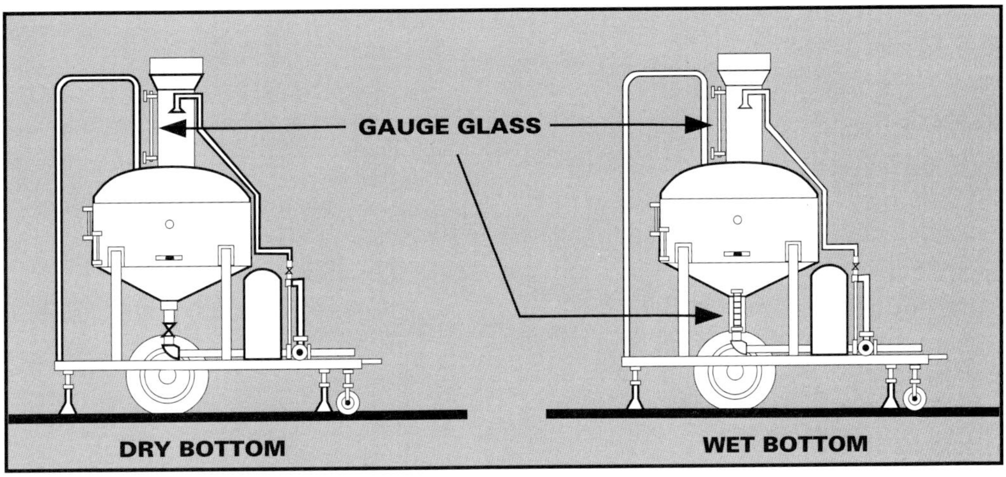

Figure 7.12. Two types of portable open-tank provers

The test liquid is the liquid run through the prover to acclimate it (i.e., to wet the tank's interior, to purge air or vapor from the system, and to stabilize the tank temperature) immediately before a proving run.

To fill a portable prover, use the same bottom-loading arm or loading spout that is used to load trucks.

As mentioned in "Requirements for Proving," the prover must have a large-enough capacity to provide proving runs of adequate duration. It must be large enough to hold at least the amount of liquid delivered through the meter in a 1-minute run during normal operation. That is, if a 1-minute run at the normal flow rate delivers 1,000 gallons, then the prover must be able to hold at least 1,000 gallons. It is preferable that the prover hold 1½ to 2 times the volume delivered during a 1-minute run: 1,500 to 2,000 gallons in this case.

A cyclic calibrator is added to some displacement meters to correct the meter's registrations every few gallons. This device adjusts the meter's drive train in fine increments to register true volume. Without a calibrator, a meter can typically be geared only to register true volume within 0.5 to 1.0%. If the meter is equipped with a calibrator, the prover must be sized in multiples of the calibrator's correction increment. For example, if the calibrator makes these incremental adjustments every 5 gallons, then the prover volume should be a multiple of 5 gallons, say, 50 to 55 gallons. If the prover volume is not a multiple of 5 gallons, then the same number of adjustments may not occur during each proving run, and the meter will fail to repeat within acceptable limits of accuracy.

Preparations for Proving with an Open-Tank Prover

Before you use any prover, take the following steps to make sure that it will perform accurately:

1. If your prover is equipped with a grounding wire to prevent static discharge during proving, attach the wire to a conducting metal (e.g., steel) that is firmly in contact with the ground.
2. Open the hatch on the neck and inspect the inside of the prover for any foreign matter, rust, or buildup on the prover walls. You may need an explosion-proof flashlight. Buildup occupies space intended for the liquid being measured and can cause inaccurate readings. If necessary, clean the prover with an appropriate solvent before proceeding. If the prover has an internal coating, make sure the coating is not peeling or damaged in any way. If it is damaged, the prover should not be used.
3. Check the body of the prover to make sure it is not dented. Dents could affect the amount the prover holds.
4. Inspect the graduated gauge glasses to be sure they are securely positioned, sealed, and clean.
5. Attach the loading arm or spout and vapor-recovery system hose, if available. If the meter is equipped with an automatic set-stop, set it to deliver a volume the prover will hold. Just be ready to shut off the flow manually if necessary. When filling the tank, watch the gauge glass. If the liquid reaches the top of the glass, shut off the flow to prevent overfilling the prover.
6. Adjust the leveling jacks on each corner of the prover to make sure the prover is level. The prover has a built-in level for reference.
7. With the meter running at its normal flow rate, fill the prover with the liquid being proved to acclimate it. If you are operating the system manually, stop the flow when the level is at about the middle of the top-neck sight glass. To do this, slow the flow when the liquid reaches the upper neck and then gradually but continuously bring the level up into the sight glass. If the meter has a set-stop, be ready to shut off the flow if the liquid reaches the top of the glass. Make sure that you operate the system smoothly. Sudden starts and stops cause undue stress on the meter and can damage it.

8. Once the prover is filled, check all connections, valves, and gauge glasses for leaks. Any leak must be repaired before a proving run begins. Leakage causes the volume contained in the tank to be lower than that which passed through the meter. Leak-detection devices on the drain valve and system help detect some leaks, but also be sure to check other areas, like the structural seams, gauge-glass assemblies, and prover connections for small leaks.
9. Check to make sure the prover is still level and that the weight of the loaded prover is not on the tires. If it is, the tank may not be completely level and, as a result, give inaccurate readings.
10. Drain the prover. With most dry-bottom provers, you can observe the drain valve. When the liquid no longer flows in a steady stream, that is, when the flow ceases and dripping starts, give the liquid clinging to the walls of the tank time to slide off and flow out of the drain. If the prover is using some type of hose or pipe connected to the drain valve, use the calibration report accompanying the prover and check the time listed for specific amounts to drain from this particular prover. As it drains, listen carefully and you can usually hear when the flow ceases and dripping begins. Allow the drain to drip for at least 30 seconds, the time it takes water to drain from the walls of the tank. (This time period is established when the prover tank volume is calibrated.) Record the drip time and allow the same amount of drip time during each run. If you are draining a wet-bottom prover, close the drain when the liquid reaches about the middle of the bottom scale. Allow calibration drip time and then record the reading on the bottom scale. You must record the liquid level so that you accurately measure the amount occupying the prover during the subsequent proving runs. Also record your drip time. Again, it is important that you allow the same amount of drip time each time you drain the prover. Be consistent with each proving run so that your volume recordings are accurate.
11. If your prover has a top spray, make sure the close-off valve is completely closed and holding before drip down begins. If not, you will have additional liquids dripping into the tank and subsequent proving runs will not be accurate.

Proving

Before starting the first proving run, look at the report form. Figure 7.9 is a copy of a typical, completed form. Different companies' forms may not be identical to this one, but most require the same information to complete the document. Make sure you complete the document accurately. Most forms have spaces for such data as the location, tender number, the type of liquid, and its API gravity. They should also have spaces for such prover data as its nominal volume at 60°F and 0 psig, and its serial number. The report should have spaces for data from previous reports, including the flow rate and the meter factor. Both of these items are necessary for comparison with previous runs and meter evaluation. Make sure you fill out all spaces, including those on meter data—its serial number, manufacturer, size, flow rate, and so on. In this example, a section for prover tank volume data is provided as well as a section for proved meter data. These sections are filled in when the proving runs are made. This report includes space for data from four proving runs; other reports may have space for more runs. In the proved meter data section, note that there are also spaces for initial and final meter readings and for indicated volume by the meter. At the bottom of the report are the spaces for signatures, date, and the companies represented.

Before your first proving run, make sure the top of the chart is filled in, including the API gravity of the liquid. If it has not been determined for you, make sure you determine the gravity before going on with the proving job, because gravity has a bearing on the accuracy of the job. If the prover has a bottom gauge glass, record the reading. If possible, reset the meter to 0 and write 0 on the report as the initial meter reading for the first proving run. If the meter cannot be set to 0, record the exact meter reading, noted to the nearest 0.01 gallon or 0.001 barrel, then fill the prover tank either by manually opening and closing the fill valve or by using the automatic set-stop system. Remember to watch for leaks throughout the proving run and to operate the meter at its normal flow rate. Also check the gauge glasses periodically for evidence of entrained air or vapor, which shows up as bubbles in the liquid in the glass, because entrained air or vapor affects the accuracy of the readings.

When the prover is filled, check again to make sure it is level. Then read the exact volume of liquid in the prover to the nearest 0.01 gallon or 0.001 barrel. With clear liquids, take the reading at the bottom of the meniscus; with opaque liquids, read the top of the meniscus. Record the reading on the report under the space labeled Run No. 1. With the prover full, read the prover thermometers, average their readings, and record the average as the tank temperature to the nearest ½°F. Finally, read and record the final meter reading to the nearest 0.01 gallon or 0.001 barrel.

Before you can compare the volume delivered by the meter with the volume in the prover, you must convert both volumes to standard temperature and pressure. In order to convert these volumes, you must determine the correction factors. Take the average temperature of the tank steel—54°F, for example—and locate that temperature in the tables of correction factors found in API *MPMS*, chapter 12, "Calculation of Petroleum Quantities." Table 7.1 reproduces the table and shows that, since 54°F falls between 52.0 and 57.3°F, the factor is 0.9999. This figure (C_{ts}) should be recorded on line 3 of the report, under prover tank volume data. Using other tables in API *MPMS*, chapter 12, section 2, you find and record the correction factor for temperature of the liquid (C_{tl}) and record it on line 4, under prover tank volume data.

Since open-tank provers are not under pressure, you do not need to determine the pressure correction factor for the prover. If the meter's operating pressure and the prover's operating pressure are the same and they remain the same from one proving run to the next, you do not need to find a pressure correction factor for the meter. At most truck-loading racks, the prover operates at atmospheric pressure. Even though the meter operates at pump pressure of 20 to 30 psig, the pump pressure is so low that the compressibility of the liquid is insignificant.

Some meters have a temperature compensator, which automatically corrects the meter to register the liquid as though it were 60°F. If the meter has a compensator, the C_{tl} factor at the meter is 1. If the meter is not temperature-compensated, you read the meter thermometer, check the correction factor tables, find and then record the C_{tl} on line 4 of prover tank volume data.

Once you have the data from the proving run and the correction factors recorded, you can convert prover volume and meter volume to a standard temperature and pressure and compare them. First, multiply the C_{ts} by the C_{tl} and record the product rounded to four decimal places on line 5 of the report. Next, multiply the delivery to the tank (the amount recorded on line 1) by the combined correction factor (the number recorded on line 5). Record the corrected prover volume (properly rounded) on line 6 of the report. Now find the corrected meter volume by multiplying the indicated meter volume by the meter's C_{tl}. Record the corrected meter volume on line 16.

Some meters at truck-loading racks are mechanically adjusted, since they deliver relatively small amounts of liquid at one time. If the meter is to be mechanically adjusted, calculate the percentage of accuracy for the meter by dividing the corrected prover volume by the corrected meter volume. For example, if the corrected prover volume is 1,002.45 and the corrected meter volume is 1004.22, then the quotient, or meter factor, is 0.99823. Record the factor on line 17 under Run No. 1. Make at least three runs with the tank prover to ensure accuracy. Add the meter factors for the three runs and divide the sum by three to find the average. Record this quotient in the box labeled Meter Factor (Average Value).

If the meter fails to repeat within the required accuracy, or the change in meter factor since the last proving exceeds tolerance, the meter should be inspected. Before recommending repair or replacement, however, make sure the proving operation was performed accurately and that the meter factor is being compared to one developed at the same flowing conditions. Was the flow rate the same for each run? Did the API gravity remain constant? Was the liquid in the system free of trapped air and vapor? Was the system free of leaks? Did the prover walls remain free of buildup? Was the prover size adequate? Was the prover level? Were all temperature and pressure corrections made? Are your calculations correct? If a downspout was used, was it completely drained into the prover after each run? If the proving operation was performed correctly, then the meter may need repair or replacement.

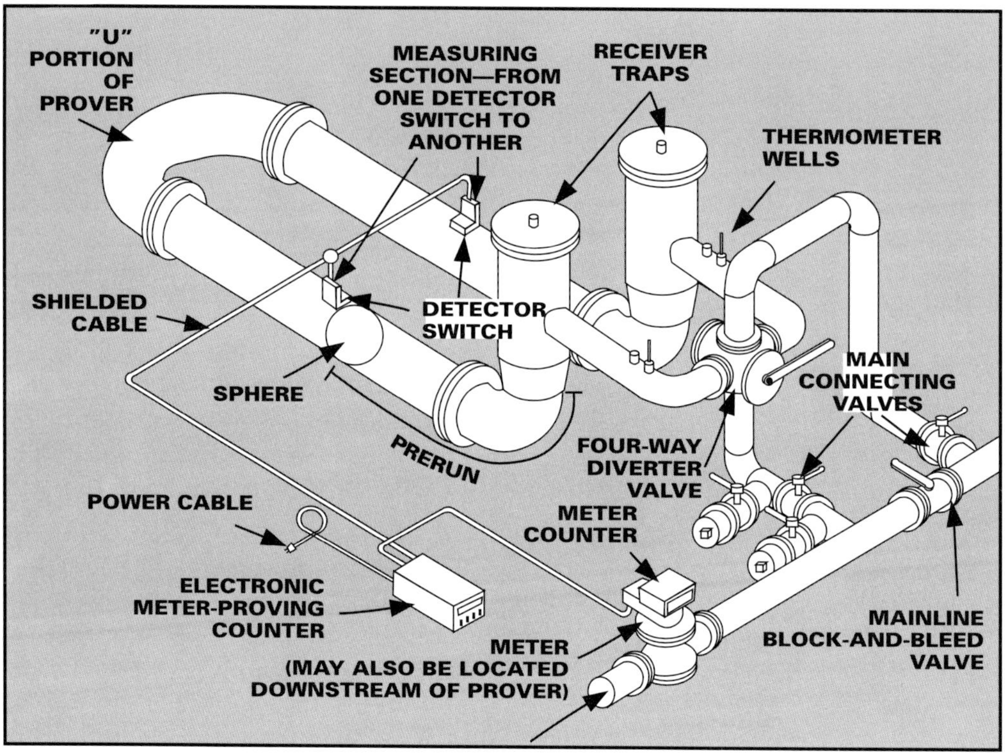

Figure 7.13. Bidirectional U-type pipe prover

PROVING METERS WITH PIPE PROVERS

Proving meters with pipe provers is somewhat different from proving with open-tank provers. There are many types of pipe provers and each type operates differently. Since bidirectional U-type pipe provers are very common, however, this type will be used to give an overview of the proving process with pipe provers.

Basic Operation of Bidirectional U-Type Provers

Basically, a bidirectional U-type pipe prover consists of several valves and connected pipe (fig. 7.13). The U-portion, or loop, as it's called, at one end gives the prover its name.

The portable U-type pipe prover is transported from site to site on the back of a truck. There it is connected to the mainline—the line carrying the liquid being metered—through connecting valves. The prover can be installed either upstream or downstream from the meter. During normal operations, the liquid flows past these connecting valves through the main valve down the mainline. During a proving operation, the main valve is closed and the liquid diverted through the prover before flowing down the mainline. Even though the liquid is diverted from the mainline, it always flows through the meter.

The four-way diverter valve on the prover controls the direction of flow through the prover. In figure 7.14 the valve is positioned so that flow goes to the left side of the prover, in a clockwise direction.

The valve is not always in this position. It can be repositioned to direct the flow counterclockwise, thereby making the prover bidirectional.

In addition to the four-way valve, the prover has two detector switches, which are connected to an electronic meter-proving counter with electric cable. The counter is connected to a pulse generator on the meter. The pulse generator puts out a large number of pulses for every gallon or barrel that the meter registers. For example, one pulse generator may put out 16,800 pulses for every barrel it measures. A high number of pulses per barrel or gallon breaks down the meter's volume measurement into very small increments. This allows a much finer and more accurate measure of the meter's indicated volume. How these pulses are used is explained below.

Between the detector switches lies the accurately calibrated part of the prover called the measuring section (see figure 7.13). This section is certified to deliver a very specific round-trip volume of liquid. That is, the maximum volume of liquid the prover can contain in this portion is certified to be a specific amount at 60°F and 0 psig—10.02806 barrels, for example. Immediately preceding each detector switch is a receiver trap and prerun. The trap is designed to hold a sphere, or ball, which is normally inside the prover. Before a proving run begins, the sphere is placed in one of the receiving traps. Liquid flow is diverted to the side of the prover holding the sphere and forces it through the prerun and toward the measuring section. As the sphere moves past the detector switch, a signal from the meter is sent to the electronic counter via a shielded cable. The counter begins to register the pulses put out by the pulse generator on the meter. As the sphere travels through the measuring section, the electronic counter continues to operate and measure the

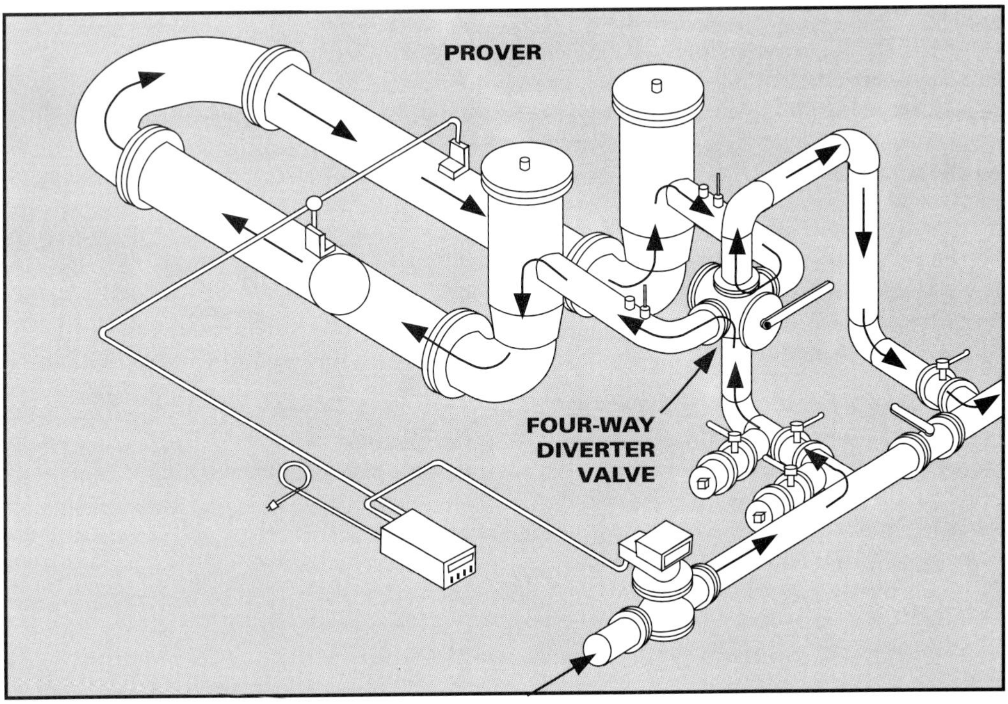

Figure 7.14. Four-way diverter valve

meter's indicated volume by counting the pulses. When the sphere reaches the other detector switch, it trips it and the counter stops registering pulses. The sphere goes into the opposite receiver trap. Half of the proving run is complete.

To finish the run, the four-way valve is turned to force the liquid and thus the sphere back through the prover. Again, the sphere actuates the detector switch, and the counter starts counting meter pulses, adding to the number already counted in the first half of the run. Finally, the sphere travels past the other detector switch and the counter stops registering pulses. One proving run in a bidirectional pipe prover is then completed. The total number of pulses for the round trip is recorded and converted to gallons or barrels to indicate the volume the meter registered. This amount, once converted to standard temperature and pressure, is compared with the certified volume of the prover's measuring section.

Just as with the open-tank prover, more than one proving run is required to make sure the results are accurate. The number of runs required depends on the policies of the involved companies. The main idea is to make runs to ensure repeatability, that is, to make runs until a certain number of pulse counts falls within acceptable tolerance. Figure 7.15 is a sample report with ten runs. Requirements called for five consecutive runs out of a maximum of ten to be the same or very close to the same. In this case, the last five are acceptable.

Preparations for Proving with a Pipe Prover

Figure 7.15 is a copy of a typical proving report for use with pipe provers. Different companies may have their own reports, but the one shown here reflects the information found on most reports. Notice that it is similar to the report used with open-tank provers, with a few exceptions. At the top of the report are spaces for the date, location of the meter, ambient temperature, prover and meter data, flow rate of liquid, and data from the previous proving. The middle has spaces for the temperature, pressure, and number of pulses for several proving runs. The lower part of the report provides space for the API gravity and type of liquid and for field calculations. The top of the report should be filled out before the first run is made and the rest as the runs are completed.

During a typical proving run with a portable bidirectional U-type prover, the first thing you do is attach a grounding cable from the prover to the meter skid. Grounding prevents the buildup of any static electricity that could spark and cause a fire. Next, remove the cap from the prover connections on the meter facility and attach the hoses from the prover to the connections on the meter station. Be careful not to spill any oil on the ground. Once the hoses are secure, open the prover inlet and outlet valves and close the mainline block valve. Allow the oil to flow through the system while continuing with the preparations. Check for any leaks in the system. If there are any leaks, the proving is not accurate.

GENERAL PURPOSE METER PROVING REPORT FOR USE WITH PIPE PROVERS

LOCATION	DATE	AMBIENT TEMP.	REPORT NO.
Lease X	9/18	76°F	1

PROVER DATA

BASE VOLUME AT 60°F AND "0" psi.	SIZE	WALL
5.0991 bbl.	10"	0.375

PREVIOUS REPORT

FLOW RATE	FACTOR	DATE
900 bbl/hr.	0.9976	8/15

METER DATA

SERIAL NO.	METER NO.	PULSES/bbl.	TEMP. COMP.	MANUF.	SIZE	MODEL
12345	1	16,800	Yes	X	6"	ABC

FLOW RATE	NON-RESET TOTALIZER
900 bbl/hr.	

RUN DATA

TEMPERATURE PROVER AVG.	TEMPERATURE METER	PRESSURE PROVER	PRESSURE METER	TOTAL PULSES	RUN NO.
72	TC	88	92	~~85,402~~	1
72	TC	88	92	~~85,312~~	2
72	TC	88	92	~~85,357~~	3
72	TC	88	92	~~85,360~~	4
72	TC	88	92	~~85,422~~	5
72	TC	88	92	85,360 ✓	6
72	TC	88	92	85,357 ✓	7
72	TC	88	92	85,359 ✓	8
72	TC	88	92	85,360 ✓	9
72	TC	88	92	85,359 ✓	10
72	TC	88	92	85,359	AVG.

C_{ts} = CORRECTION FOR TEMPERATURE ON STEEL

C_{ps} = CORRECTION FOR PRESSURE ON STEEL

C_{tl} = CORRECTION FOR TEMPERATURE ON LIQUID TABLE 6 OR TABLE 24 FOR LPG s

C_{pl} = CORRECTION FOR PRESSURE ON LIQUID

LIQUID DATA

TYPE	API GRAVITY	SPECIFIC GRAVITY	R.V. PRESS	BATCH/TENDER NO.
OIL	39.0 AT 60°F	— AT 60°F	—	

FIELD CALCULATIONS

PROVER VOLUME		C_{ts}		C_{ps}		C_{tl}		C_{pl}		CORRECTED PROVER VOLUME
5.0991	X	1.0003	X	1.0001	X	0.9940	X	1.0005	=	5.0731

AVERAGE PULSES		PULSES/bbl.		GROSS METER VOL.		C_{tl} USE ONLY FOR NONTEMP. COMP. METER		C_{pl}		CORRECTED METER VOLUME
85359	÷	16,800	=	5.0809	X	TC	X	1.0005	=	5.0834

CORRECTED PROVER VOLUME		CORRECTED METER VOLUME		METER FACTOR		C_{pl} LIQUID CORR. FOR PRESS. AT METERING COND.		COMPOSITE FACTOR USE FOR CONSTANT PRESSURE APPLICATIONS
	÷		=	0.9980	X	1.0005	=	0.9985

REMARKS, REPAIRS, ADJUSTMENTS, ETC.,

SIGNATURE	DATE	COMPANY REPRESENTED
R K Davis	9/18	Ace Oil
B L Smith	9/18	Device Distributing

(API MPMS Chapter 12.2) 29

Figure 7.15. General-purpose meter proving report for use with pipe provers

Figure 7.16. Attaching pulse generator to meter register

Next, attach the pulse generator to the meter (fig. 7.16). Connect one end of a lead to the pulse generator and the other end of the lead to the electronic pulse counter. Then connect the leads from the detector switches to the pulse counter. Test the counter to make sure it is working properly. Normally, testing means following a test sequence provided by the manufacturer.

After testing the counter, bleed off any air or vapor that may be trapped in the prover. Bleed valves are usually provided on the prover. Bleed vapor before and after each proving run. Air or vapor takes up space in the prover that should be occupied by the liquid and can cause erroneous results. Next, to make sure the flow rate, temperature, and pressure are stabilized, make 2 or 3 trial proving runs. Follow the procedures described in the following section except for recording any data. Once the prover is stabilized, you are ready to make the first run where data is recorded.

Proving the Meter

Flip the switch to actuate the four-way valve and launch the sphere. While the sphere is moving through the prover, determine the flow rate through the system. Flow rate affects the amount of slippage through the meter and different flow rates may have different meter factors. Record the flow rate on the proving report and then read the temperature of the oil at the prover's inlet and outlet thermometers. If the two readings are not within 2°F, average them. After recording the temperature, read the pressure at the meter. If both inlet and outlet pressures are available, average the two. If the meter is not temperature-compensated, read and record the line temperature at the meter during the proving run. If the meter is temperature-compensated, just write "TC" (temperature-compensated) in the space provided for meter.

Next, take a sample of the oil to measure its API gravity. Withdraw the liquid upstream or downstream from the meter and prover, since the proving run is in progress. During a proving run, the sample should never be taken between the meter and the prover. All liquid that goes through the meter must go through the prover.

After obtaining the sample, measure its API gravity and temperature. (See chapter 3 for explanations of how to measure the API gravity.) Then, using an API conversion table (table 7.2), correct the gravity to the standard temperature of 60°F. All gravity readings must be corrected to standard temperature. Record the corrected API gravity on the proving report. By this time, the sphere should have finished its first trip through the prover—half the proving run is complete. Turn the four-way valve to send the sphere back the other way to complete the run.

At the end of the run, read and record the total number of pulses registered on the counter. For example, on the proving report in figure 7.15, 85,402 pulses were registered for the first run. After recording the number of pulses, make several additional runs until the required number of pulse count totals fall within the required tolerance. (During the last run, take another liquid sample and measure the API gravity and temperature to make sure they did not change during the proving operation.) In the example given in figure 7.15, ten runs were made before five consecutive runs were acceptable.

TABLE 7.2
Generalized Crude Oils, API Correction to 60°F

	API Gravity at Observed Temperature											
Temp F	70.0	70.5	71.0	71.5	72.0	72.5	73.0	73.5	74.0	74.5	75.0	Temp F
	Corresponding API Gravity at 60°F											
75.0	68.0	68.5	68.9	69.4	69.9	70.4	70.9	71.4	71.9	72.3	72.8	75.0
75.5	67.9	68.4	68.9	69.4	69.9	70.3	70.8	71.3	71.8	72.3	72.8	75.5
76.0	67.8	68.3	68.7	69.2	69.7	70.2	70.7	71.2	71.6	72.1	72.6	76.0
76.5	67.8	68.3	68.7	69.2	69.7	70.2	70.7	71.2	71.6	72.1	72.6	76.5
77.0	67.7	68.2	68.7	69.2	69.6	70.1	70.6	71.1	71.6	72.1	72.5	77.0
77.5	67.6	68.1	68.6	69.1	69.6	70.1	70.5	71.0	71.5	72.0	72.5	77.5
78.0	67.6	68.1	68.5	69.0	69.5	70.0	70.5	71.0	71.4	71.9	72.4	78.0
78.5	67.5	68.0	68.5	69.0	69.4	69.9	70.4	70.9	71.4	71.8	72.3	78.5
79.0	67.5	67.9	68.4	68.9	69.4	69.9	70.3	70.8	71.3	71.8	72.3	79.0
79.5	67.4	67.9	68.3	68.8	69.3	69.8	70.3	70.7	71.2	71.7	72.2	79.5
80.0	67.3	67.8	68.3	68.8	69.2	69.7	70.2	70.7	71.2	71.6	72.1	80.0
80.5	67.3	67.7	68.2	68.7	69.2	69.7	70.1	70.6	71.1	71.6	72.0	80.5
81.0	67.2	67.7	68.1	68.6	69.1	69.6	70.1	70.5	71.0	71.5	72.0	81.0
81.5	67.1	67.6	68.1	68.6	69.0	69.5	70.0	70.5	71.0	71.4	71.9	81.5
82.0	67.1	67.5	68.0	68.5	69.0	69.5	69.9	70.4	70.9	71.4	71.8	82.0
82.5	67.0	67.5	67.9	68.4	68.9	69.4	68.9	69.4	69.9	70.3	71.8	82.5
83.0	66.9	67.4	67.9	68.4	68.8	69.3	69.8	70.3	70.7	71.2	71.7	83.0
83.5	66.9	67.3	67.8	68.3	68.8	69.2	69.7	70.2	70.7	71.2	71.6	83.5
84.0	66.8	67.3	67.8	68.2	68.7	69.2	69.7	70.1	70.6	71.1	71.6	84.0
84.5	66.7	67.2	67.7	68.2	68.6	69.1	69.6	70.1	70.5	71.0	71.5	84.5
85.0	66.7	67.1	67.6	68.1	68.6	69.0	69.5	70.0	70.5	70.9	71.4	85.0
85.5	66.6	67.1	67.6	68.0	68.5	69.0	69.5	69.9	70.4	70.9	71.4	85.5
86.0	66.5	67.0	67.5	68.0	68.4	68.9	69.4	69.9	70.3	70.8	71.3	86.0
86.5	66.5	67.0	67.4	67.9	68.4	68.8	69.3	69.8	70.3	70.7	71.2	86.5
87.0	66.4	66.9	67.4	67.8	68.3	68.8	69.3	69.7	70.2	70.7	71.1	87.0
87.5	66.3	66.8	67.3	67.8	68.2	68.7	69.2	69.7	70.1	70.6	71.1	87.5
88.0	66.3	66.8	67.2	67.7	68.2	68.6	69.1	69.6	70.1	70.5	71.0	88.0
88.5	66.2	66.7	67.2	67.6	68.1	68.6	69.1	69.5	70.0	70.5	70.9	88.5
89.0	66.2	66.6	67.1	67.6	68.0	68.5	69.0	69.5	69.9	70.4	70.9	89.0
89.5	66.1	66.6	67.0	67.5	68.0	68.4	68.9	69.4	69.9	70.3	70.8	89.5
90.0	66.0	66.5	67.0	67.4	67.9	68.4	68.8	69.3	69.8	70.3	70.7	90.0

Source: API, Standard 2540, Petroleum Measurement Tables, Volume Correction Factors, vol. I, table 5A.

*Notes:*API gravity = 70.0–75.0. All values are extrapolated.

Once you have a sufficient number to establish repeatability, cross off those that do not meet the requirements and average the total for the acceptable runs. In figure 7.15, the last five pulse counts would be averaged. Next, average the temperature and pressure readings for both meter and prover and record them in the spaces provided. To determine the gross meter volume, divide the averaged pulse count by the number of pulses per barrel. In figure 7.15 the gross meter volume is 5.0809 barrels. Now enter the prover's measuring section in the space provided.

Just as you do with open-tank proving, you must convert both these volumes to what they would be at standard temperature and pressure. To find C_{ts}, C_{ps}, C_{tl}, and C_{pl}, use the appropriate API tables. Table 7.1 reproduces one of these tables, table A-1 from chapter 12, section 2 of API *MPMS*. This table gives you the C_{ts} values for mild steel, the material used to make many provers. Locate the observed temperature of the prover on the table and move across to the next column for the C_{ts} value. For instance, if the temperature were 78°F, the C_{ts} value would be 1.0003. This figure is recorded in the C_{ts} space provided on the form (fig. 7.15). In similar fashion and using other tables provided by the API standards, find the values for C_{ps}, C_{tl}, and C_{pl}, and enter them on the form. Multiply all of these figures together using proper sequence and round-off procedures to find the corrected prover volume—5.0731 in the example given. See API *MPMS*, chapter 12, section 2, "Calculation of Petroleum Quantities," for the rules governing the sequence and rounding of numbers.

Next you correct the gross meter volume in a similar way. If the meter is temperature-compensated, the C_{tl} for the meter is 1. You record the corrected volume in the space provided. Once both volumes have been corrected to standard temperature and pressure, you can determine the meter factor by dividing the corrected prover volume by the corrected meter volume. In this case, the meter factor—.9980—is recorded on the report.

The meter factor can now be used to correct the meter's registered volume. To correct the meter registration, multiply the registered volume by the meter factor. For example, if the meter registers 10,000 barrels, then the corrected reading would be 10,000 × .9980, or 9,980 barrels. You use this meter factor to correct the meter's register until the next proving run.

PROVING SCHEDULES

Proving schedules vary among companies. Time is not the most important factor to consider when scheduling proving runs—wear on the meter is much more important. Wear depends on the volume through the meter. Many companies prove a meter after 200,000 barrels of throughput volume (volume flowing though the meter), or once a month, whichever comes first. Others have different requirements. The following are proving schedules for two major oil companies, based on volume throughput per month:

Company A

5,000 barrels/month or less, prove quarterly
5,001–7,500 barrels/month, prove bimonthly
Greater than 7,500 barrels/month, prove monthly

Company B

5,000–15,000 barrels/month, prove bimonthly
Less than 5,000 barrels/month, prove quarterly
If throughput is small, prove every 60–90 days

CONCLUSION

Meters measure the amount of liquid flowing through them, but they cannot do this with 100% accuracy. Since even small measurement errors can cost a lot of money, all meters must be proved periodically to determine if they are in error. Though the types of meters and provers may vary from the ones explained and illustrated in this chapter, the principles and mechanics of operation are similar. Follow the procedures described here and in API *MPMS,* chapter 4, section 8, "Prover Operations," (currently being written and expected to be published in 1994), to ensure that the liquid flowing through the meter is accounted for accurately.

CHAPTER 8

Orifice Meter Installations

INTRODUCTION

Measuring petroleum liquids, as described in chapter 7, is a straightforward process, but, because of the nature of natural gas, a direct measure of its volume is very difficult. Natural gas is a mixture of hydrocarbon gases with very low density and viscosity; as a result, a volume of gas expands and contracts easily, depending on the pressure and temperature. Natural gas distributes itself uniformly throughout a container and is also invisible, odorless, and lighter than air. (The odor we associate with natural gas is a sulfur compound called mercaptan. It is added to natural gas so that any leaks can be detected.) In other words, it is a devil of a substance to contain, much less to measure, since its density and volume change dramatically with any change in pressure and temperature. There is no gallon, or barrel, for that matter, of natural gas. Natural gas must, by its nature, be contained and measured in a very different way from liquids.

BEHAVIOR OF GAS MOLECULES

Gas molecules are in constant motion. They travel in a straight line until they collide with another molecule or a confining wall. They bounce off each other much the way billiard balls do when they strike. The pressure exerted by a gas depends on how hard and how often its molecules collide with the walls of the chamber in which it is confined. The fewer the gas molecules in a chamber, the fewer the molecular collisions, and thus the lower the exerted pressure. The greater the number of gas molecules in a confined space, the greater the number of collisions, and thus, the greater the pressure exerted by the gas.

Gas molecules are always in motion, and their speed is affected by temperature. The higher the temperature, the faster the gas molecules move. Faster motion causes more collisions between the molecules and with the confining wall. As long as the volume of the container does not change, pressure increases as temperature increases. Conversely, the lower the temperature, the more slowly the gas molecules move, causing fewer collisions between molecules and with the confining walls. Under constant volume, pressure decreases as temperature decreases.

These two facts about gas molecules allow us to measure natural gas even though we cannot see it. You can use the pressure exerted by the gas and its temperature to determine how many gas molecules are within a given space. For example, if you have two containers of the same volume, made of the same material, and containing the same type of gas, you can assume that, if the pressures on the gauges are the same and the temperatures are the same, you must have the same number of gas molecules within each space (fig. 8.1). If one of the pressure gauges registers a higher pressure but the temperature is the same in both vessels, however, the higher-pressured space contains more molecules of gas. If one of the thermometers registers a higher temperature but the pressures on both gauges are the same, the warmer container has fewer molecules moving more rapidly to produce the same pressure (fig. 8.2).

Gas molecules are also incompressible. When you compress a gas, you are not compressing the molecule, you are simply bringing the gas molecules closer together. Reducing the amount of space between molecules increases the number of collisions and thus the pressure exerted by the gas.

Pressure and temperature allow you to *infer* how many molecules are within a space. You cannot measure the number

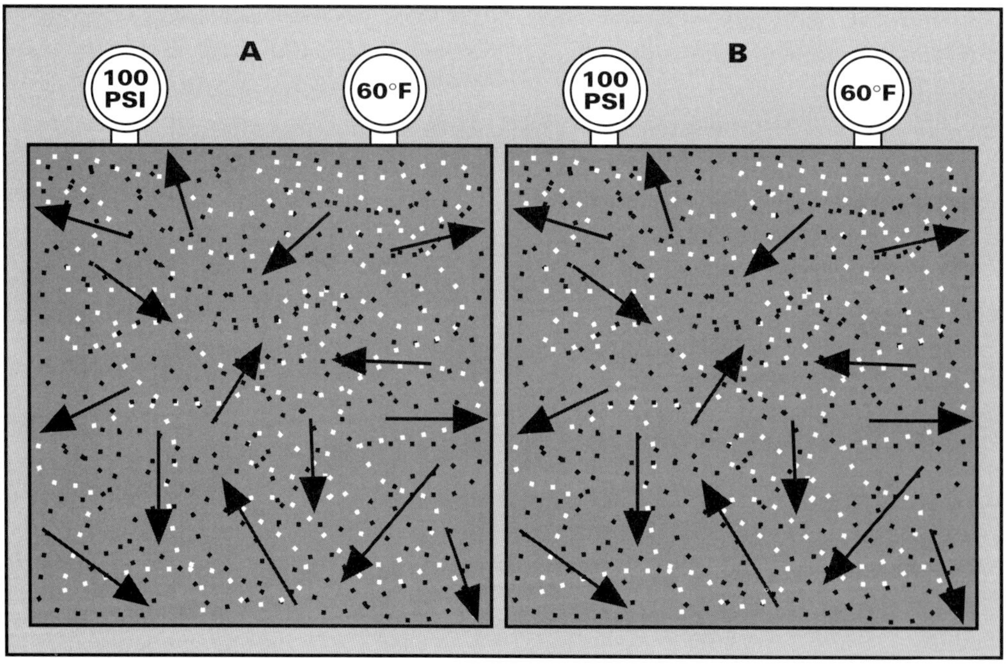

Figure 8.1. Equal-sized vessels at same pressure and temperature contain the same number of gas molecules.

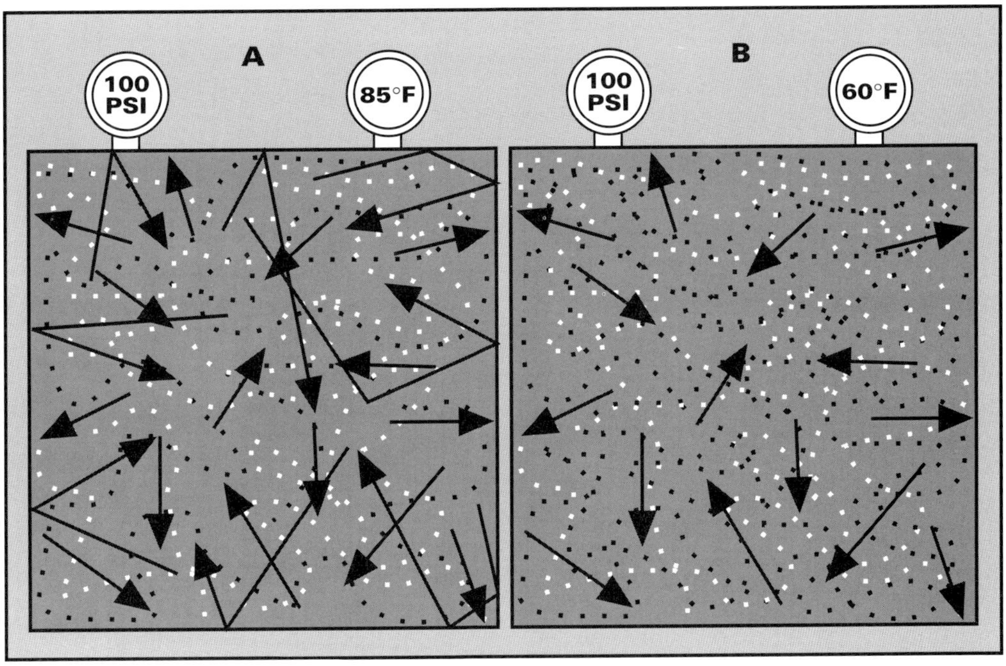

Figure 8.2. Equal-sized vessels at the same pressure but different temperatures contain different number of gas molecules.

directly but you can determine, or calculate, the number by knowing the volume of the space, the type of gas (i.e., its composition), the pressure, and the temperature of the gas.

Natural gas is stored in pressurized containers and transported under pressure through pipelines. To measure a volume of gas, you must determine how many molecules of a gas of a given density are contained within a cubic foot of space at a given pressure and temperature and then determine how many cubic feet of space the pressurized container contains, or how many cubic feet of gas have passed a particular point in a gas pipeline in a certain amount of time. The volume is normally expressed as so many cubic feet at a base temperature and pressure.

There are different ways to measure a volume of natural gas, but a widely used method is to use an orifice meter installation. The orifice meter does not give a direct readout of the amount of gas. The meter does, however, provide information that, along with mathematical formulas, allows you to calculate the amount of gas flowing through the installation and express it at base conditions of temperature and pressure. The meter is called an inferential meter since amounts are not directly measured, as in a positive-displacement meter, but are *inferred* from the differential pressure and other flowing parameters.

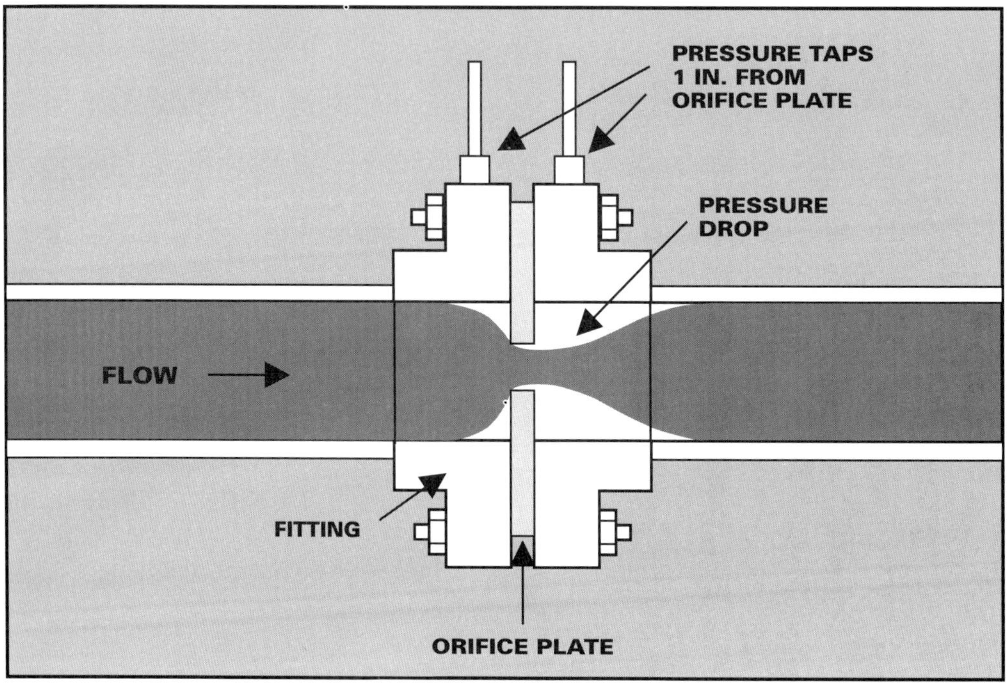

Figure 8.3. Primary element

ORIFICE METER INSTALLATIONS

Orifice meter installations consist of two basic parts: a primary element and a secondary element. The primary element (fig. 8.3) includes a section of pipe called a meter tube, an orifice plate, a fitting to hold the plate in the tube, and a pressure tap located on each side of the orifice plate. The orifice plate restricts the gas flow and reduces the flowing pressure of the gas inside the pipeline. This change in pressure is called the *pressure differential* and is recorded by the secondary element. The secondary element records the pressure change created in the primary element and other variables necessary for calculating the gas flow.

One of two means can be used to record the differential pressure, the static pressure, the temperature of the flowing gas, and the time of day—all items necessary to calculate the amount of gas flowing past a particular point in the line, in this case, the meter. On some installations the secondary element consists of a sensing element, a temperature recorder, a pressure recorder, and a chart. The differential pressure is transferred through the pressure taps and gauge lines on each side of the orifice plate to a sensing element on the recorder and is then recorded with a pen on a chart (fig. 8.4). The static pressure of the flowing gas and sometimes the temperature of the gas are also sensed and recorded on this chart. The temperature is sometimes recorded on a separate recorder and chart. On other installations, the secondary element consists of transducers that electronically convert the differential pressure, the static pressure, and the temperature to electronic signals and feed them into a computer (fig. 8.5).

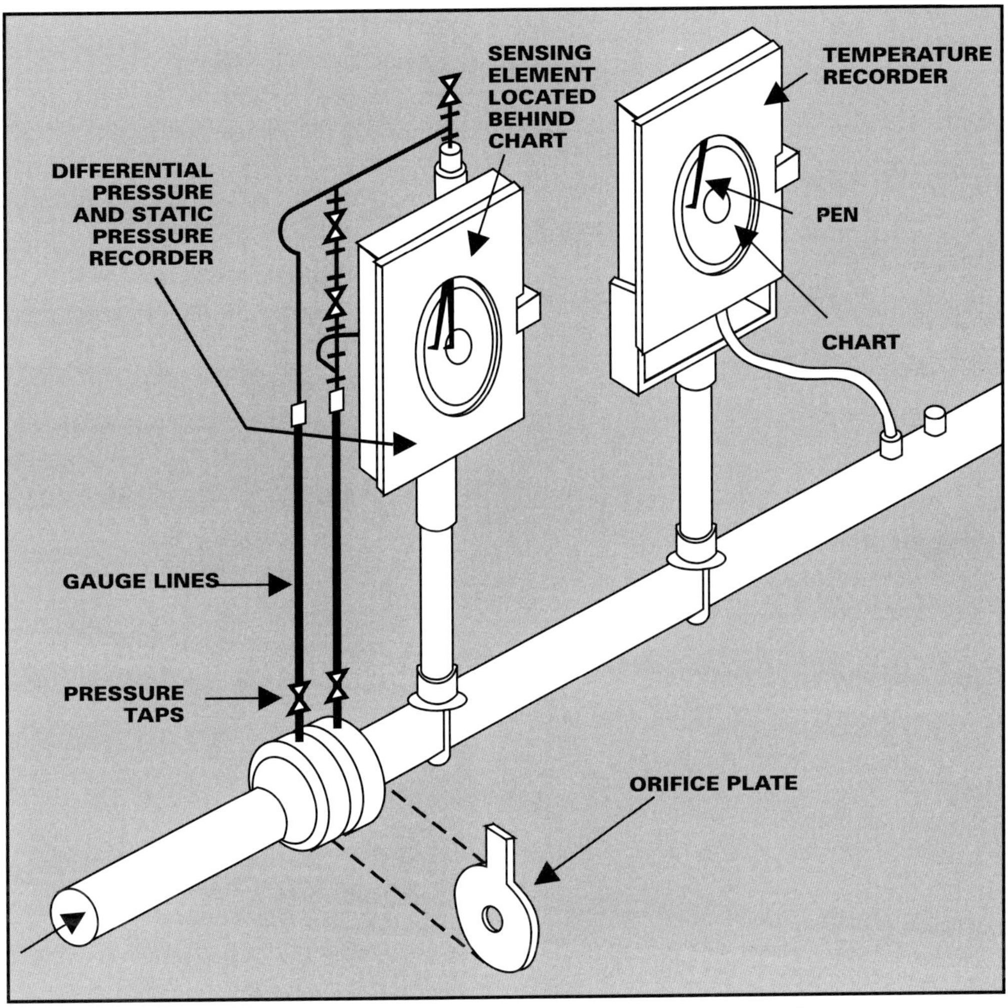

Figure 8.4. Secondary element

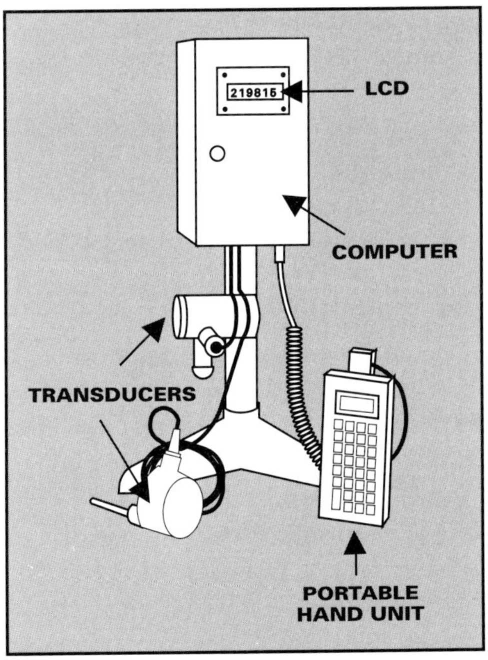

Figure 8.5. Electronic secondary element

HOW AN ORIFICE METER INSTALLATION WORKS

To understand how an orifice meter installation works, look at the heart of the installation: the orifice plate and fitting. The *orifice plate* (fig. 8.6) is a disk, usually made of stainless steel, with a precisely machined and measured hole. When the plate is placed in the line and held securely by the fitting, it constricts the flow of gas and causes a slight drop in gas pressure, which is greatest just downstream of the plate. As the gas squeezes through the orifice, its velocity increases and its pressure drops, much like what happens to water when it passes through a nozzle at the end of a hose. As mentioned earlier, this drop in gas pressure is called the pressure differential. Knowing the pressure differential across the plate and combining this information with other factors like the flowing temperature of the gas, the size of the orifice in the plate, the size of the meter tube, the static pressure in the line, and the specific gravity of the gas allows you to calculate the amount flowing through the plate—provided all parts of the orifice meter installation are built to the specifications and requirements listed in API *MPMS*, chapter 14, section 3, part 2, or the American Gas Association's *Report #3* (*AGA-3*).

Figure 8.6. Orifice plate

To explain how the differential can be used to calculate gas flow, look at the following hypothetical example. If you placed an orifice plate in a line (with gas flow at atmospheric pressure) and then attached a U-tube (a manometer) containing water to the gauge pipes on each side of the plate (fig. 8.7), the greater pressure on the upstream side of the plate would force the water down. At the same time, the reduced pressure on the downstream side of the plate would let the water on that side rise. The change in water level is a direct indicator of the pressure differential across the plate. If you attached a calibrated prover to one end of the pipeline and ran several tests, you could calculate the amount of gas flowing through the line by measuring the change in water level in the U-tube and comparing these data with the amount of gas the prover measured. The orifice meter is the indirect or inferential measure of the flow rate, the prover, the direct measure. For instance, if gas is flowing at 60°F and 35 psi through a 1-inch orifice and then through the prover, a 1-inch difference in water level in the U-tube would indicate that a certain amount of gas, say, 1,088 cubic feet/hour, was flowing through the

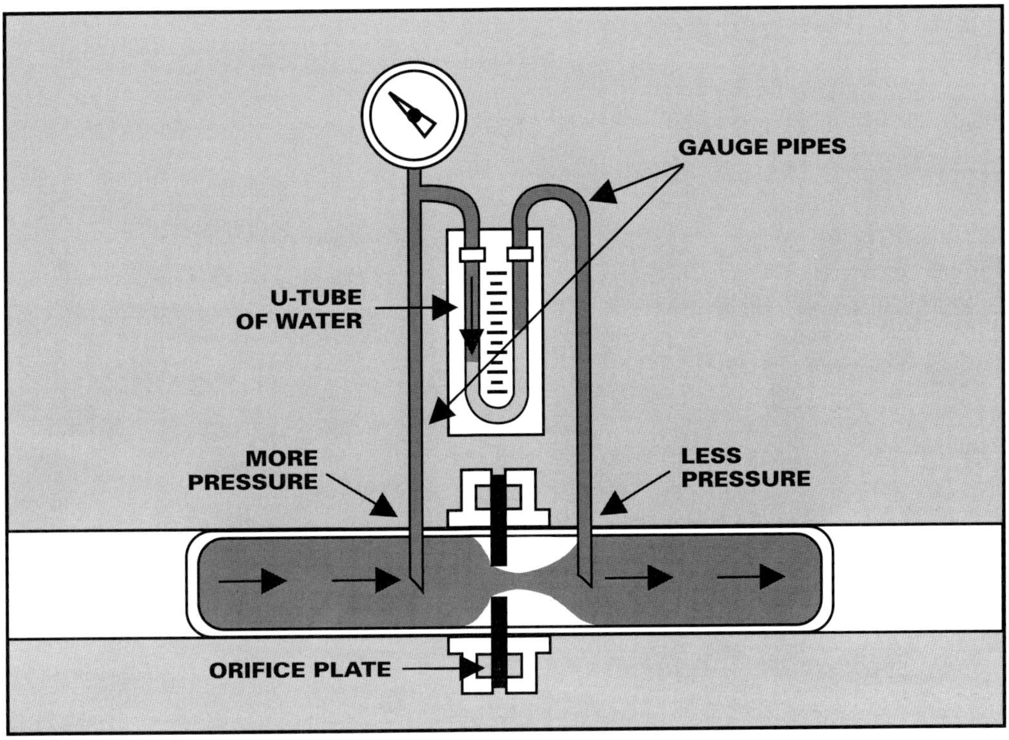

Figure 8.7. Change in water level based on pressure difference

prover. Two inches of water differential would indicate that a different amount had passed the orifice plate, 3 inches another amount, and so on. Repeated tests would show that under the same conditions certain changes in water level would mean a certain amount of gas had flowed past the plate. You would no longer need the prover; you could use the pressure differential to indicate, or to infer, the total volume. Of course, provers are not attached to pipelines in the field because all of these tests have been performed and the results published in API *MPMS*, chapter 14, section 3. These results are used to establish the calculations for determining gas flow rate from the differential reading.

But what would happen if a different-sized orifice were used on this same pipeline, or if there were a change in the temperature or the pressure of the flowing gas? The differential would change and a different flow rate would be calculated. Remember, any change in temperature or pressure affects gas measurement.

Not only field conditions affect gas measurement; certain characteristics of the gas also affect it. Any changes in these conditions would have to be taken into account by using mathematical factors and a formula that take into account any changes that might occur. The differential pressure, the static pressure, and the flowing temperature of the gas are recorded for particular time periods and then used along with the specific gravity of the gas in a somewhat complex formula (discussed later in this chapter) to determine the amount flowing past the orifice meter.

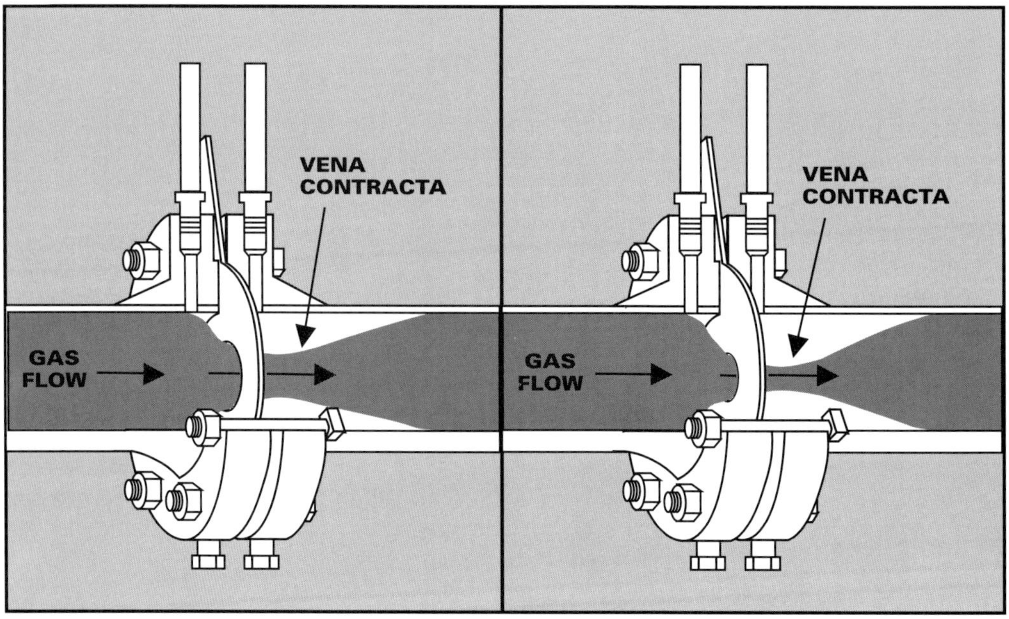

Figure 8.8. Vena contracta

THE PRIMARY ELEMENT

In order to have accurate measurement, all mechanical parts of the orifice meter station must be the correct size, be in good condition, and be built and installed according to the requirements of API *MPMS,* chapter 14, section 3.

Orifice Plates

Orifice plates, for example, come in many line diameters with different-sized orifices. Different line diameters allow you to have plates that fit into different-sized pipelines; different-sized orifices allow you to vary the rate of the gas flow as it moves past the plate. Each size produces a specific *vena contracta,* or squeezing of the gas flow (fig. 8.8).

The shape of the vena contracta is important because it affects the downstream pressure reading. An orifice is designed to produce a specific vena contracta and thus a specific pressure drop. To calculate the flow of gas through a particular installation, you must know the size of the orifice because the size is one factor in the equation for calculating gas flow. What size orifice you use depends on the flow rate of the gas. The higher the flow rate, the larger the orifice.

The thickness of the plate is also important. The plate is usually only as thick as is needed to keep it from buckling or warping under pressure. As stated earlier, each plate is designed to produce a specified vena contracta. If the plate is not the right thickness, it alters the flow of gas and produces an incorrectly shaped vena contracta, which in

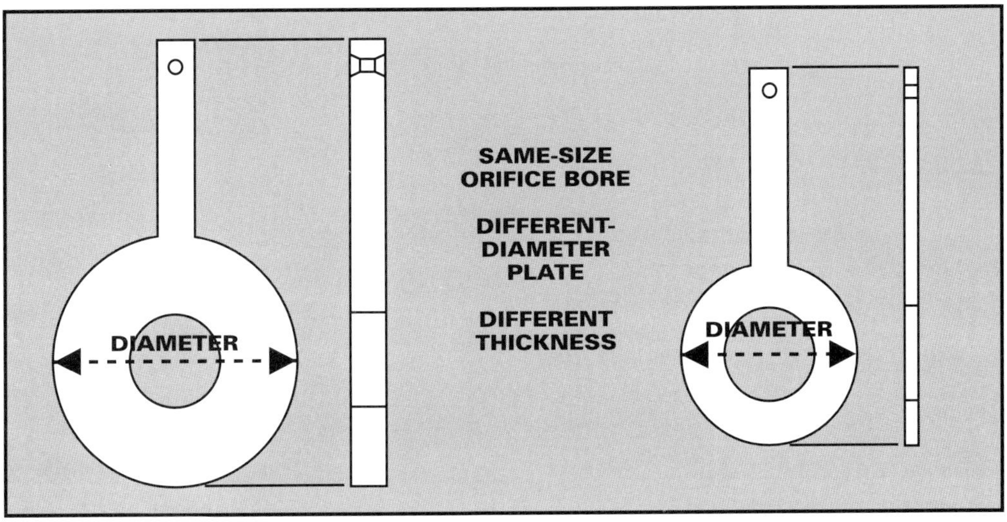

Figure 8.9. Orifice plates

turn produces a differential that is lower than it should be, and thus prevents accurate gas measurement.

Figure 8.9 shows different plates with the same-sized orifice but not the same plate thickness or diameter. Smaller line diameter plates are generally thinner than larger plates. Even so, some of the smaller plates are still too thick relative to their diameter to produce the correct vena contracta. To solve this problem, these plates are usually beveled (fig. 8.10).

Beveling allows the gas to contact the thickness only at the front, or unbeveled, edge of the plate. It then flows past the orifice and produces the correct vena contracta. The sharp, unbeveled edge of the orifice always faces upstream. Reversing it alters the vena contracta and lowers the differential 20% to 35%.

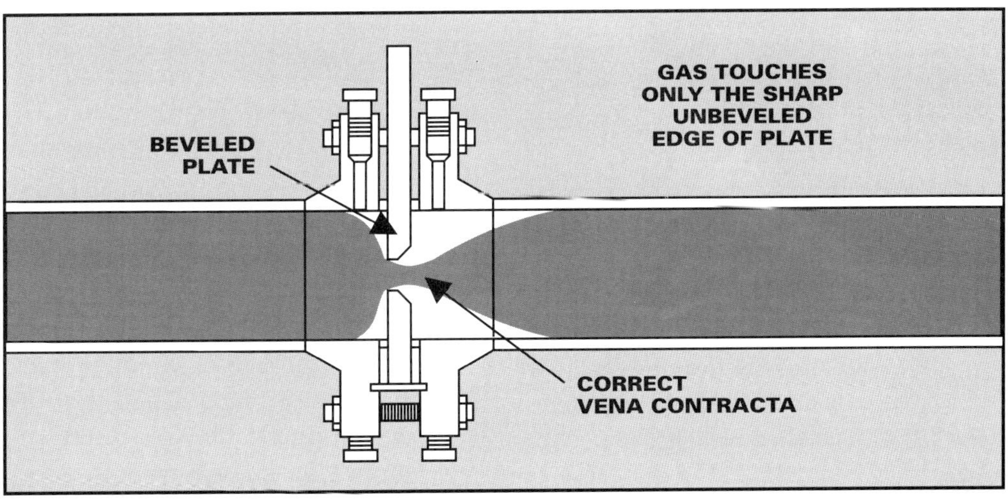

Figure 8.10. Beveled orifice plate

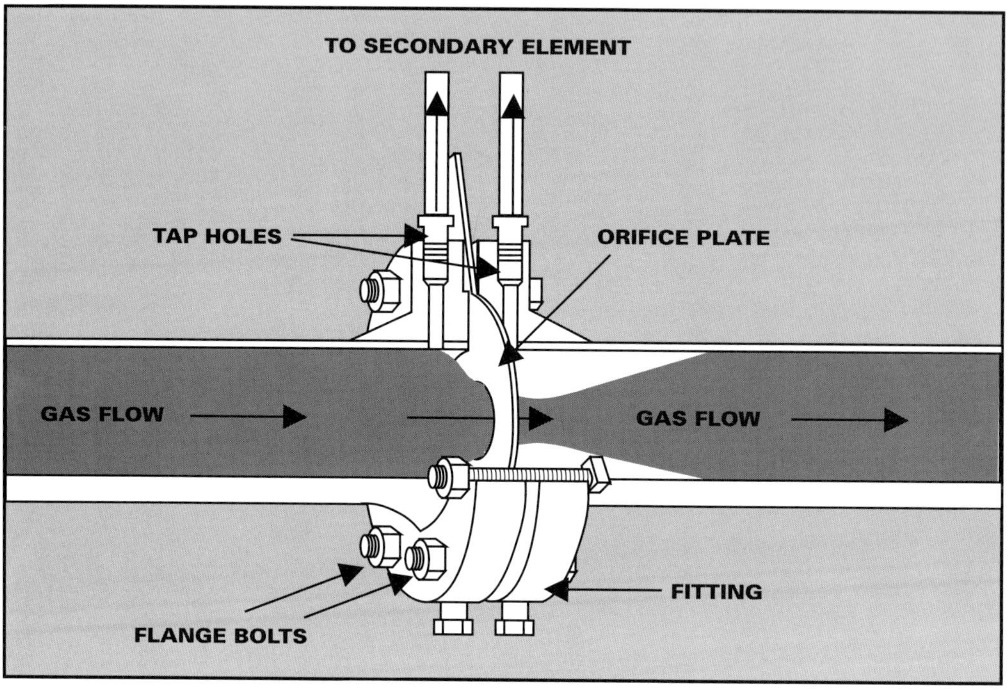

Figure 8.11. Flanged fitting

Fittings

Not only the size and the width of the plate are important, but also how carefully it is positioned in the line. The fitting holds the plate firmly in the center of the line at 90° to the gas flow so that all gas passes through the orifice. Since gas measurement is based on the pressure differential across the plate, any gas leaking around the plate and not going through the orifice will cause errors in measurement. The fitting also provides a way for the plate to be removed, inspected, and, if necessary, replaced. In addition, the fitting comes equipped with holes called taps, located 1 inch from each side of the plate. The pressure differential is sensed through these taps and recorded on the secondary element.

Different types of fittings are available. Which type you use depends on company policy, government regulations, economics, and the manufacturer preferred. Four common types are the flanged, the single-chambered, the single chamber with gearing, and the dual-chambered.

Flanged fittings (fig. 8.11) are precision-made companion flanges with tap holes. These tap holes are usually ⅜ inch in diameter for 2- and 3-inch meter runs and ½ inch in diameter for runs through 12 inches and provide a connection between the meter run and the secondary element. When installed as required by the standard, the orifice plate must be accurately centered between the flanges. The bolts must be uniformly tightened so that neither flange face is pulled out of line; pulling them out of line causes gas leaks. With a flanged fitting you must isolate the meter run (i.e., stop the gas flow through this part of the line) and bleed it down (i.e., depressurize the meter tube to atmospheric pressure) before loosening the flange bolts to remove and inspect or change the plate.

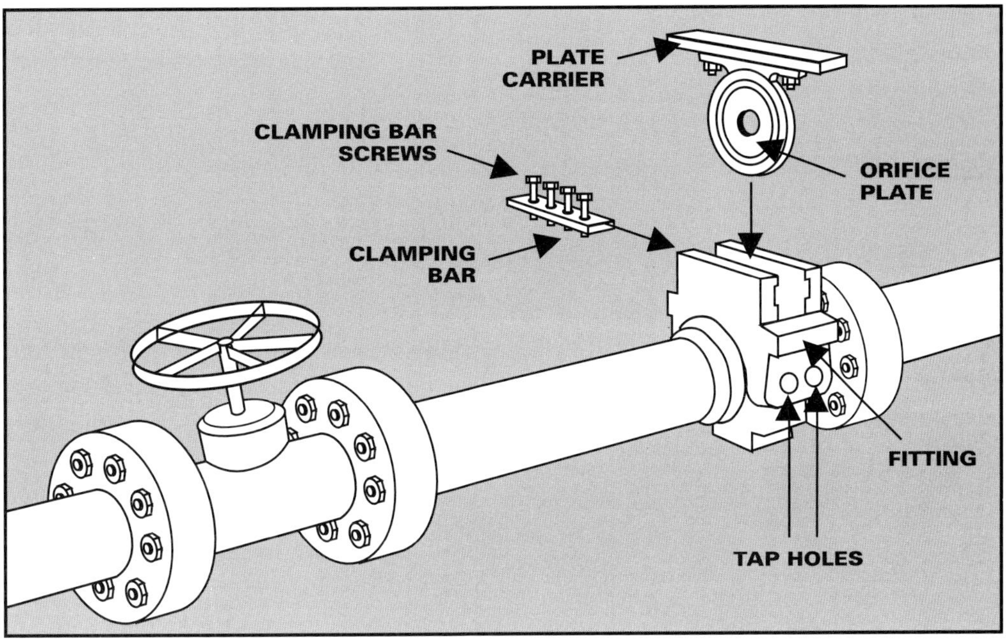

Figure 8.12. Single-chamber fitting

A single-chamber fitting (fig. 8.12) is a modified version of the flanged fitting. The flange bolts have been eliminated and the fitting has one chamber that holds the plate. To remove and inspect the plate, you isolate the meter run and bleed it down, just as you do with the flanged fitting. Then, you loosen the clamping bar set screws and slide the bar out and then lift the plate and carrier out for inspection.

The single chamber with gearing (fig. 8.13) is a modified version of the single-chamber fitting. It is equipped with a geared shaft that rolls the plate and carrier up and out for inspection. You still have to isolate the meter tube and bleed it down before removing the plate.

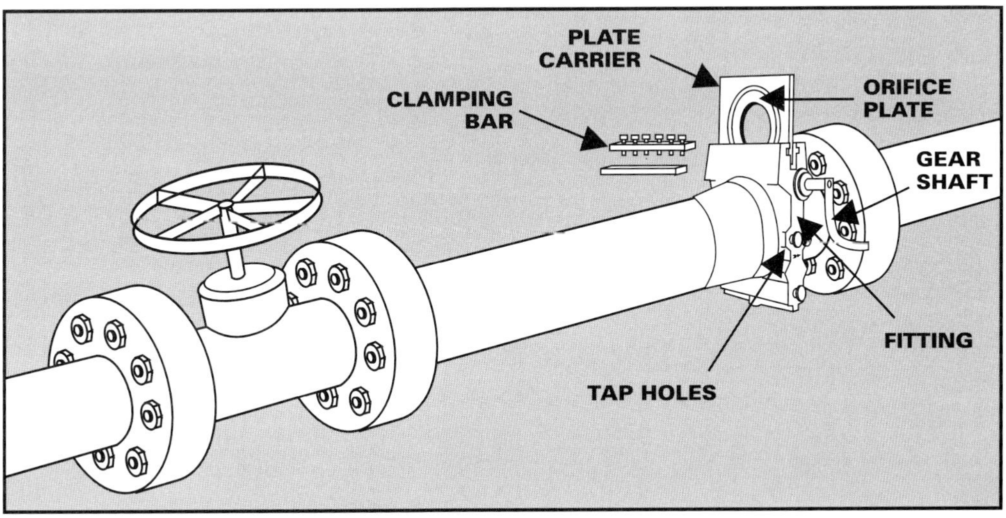

Figure 8.13. Single-chamber fitting with gearing

The dual-chamber fitting (fig. 8.14) is more complex than the others. It allows you to remove the plate and carrier without depressuring the meter run or shutting off the gas flow. To remove the plate with this fitting, you roll the plate carrier from the lower chamber to the upper, close the valve to isolate the plate from the lower chamber, bleed the gas from the upper chamber until the upper chamber is at atmospheric pressure, remove the clamping bar, and then roll the plate out of the fitting. The gas flows through the line unimpeded.

All four fittings have been illustrated in an upright position to show how they work. Meter tubes and fittings are often installed so that the tap holes are facing upward (fig. 8.15) in situations where free liquids in the gas can accumulate at the orifice plate. The taps face up so that any of these liquids will drain back into the fitting and not plug the sensing lines to the secondary element. As stated earlier, the differential pressure is sensed through these taps and any liquids plugging the tap holes or the sensing lines will affect the differential pressure reading.

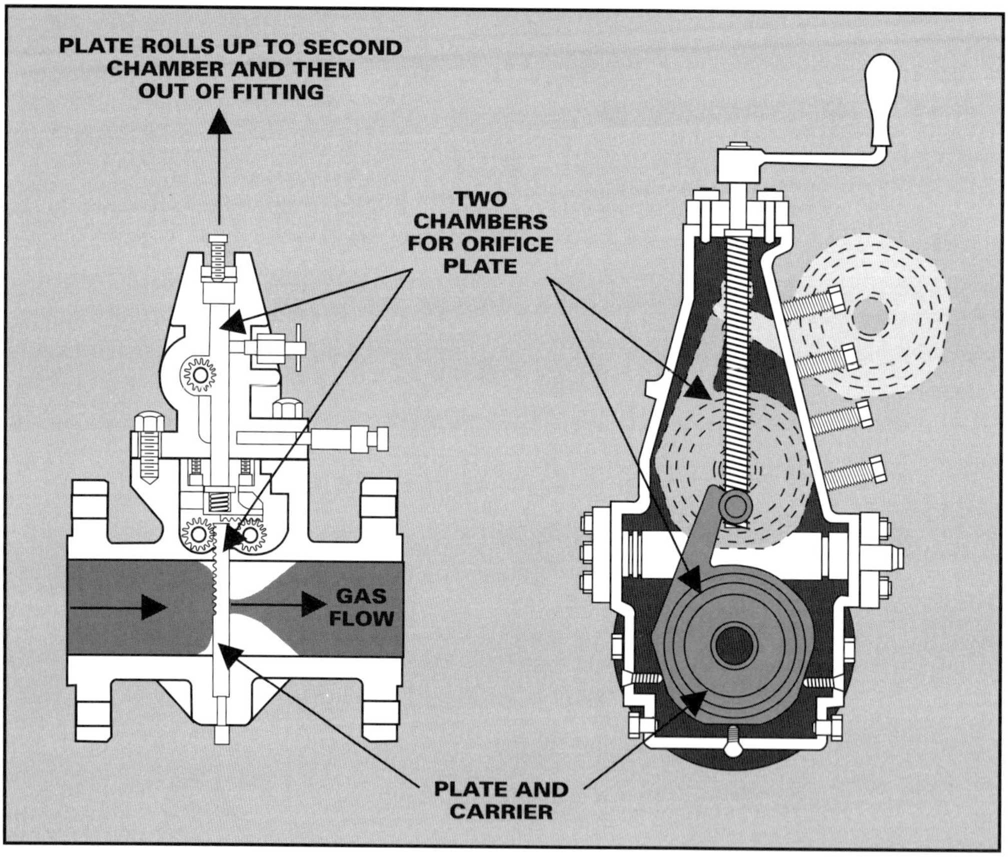

Figure 8.14. Two types of dual-chamber fitting

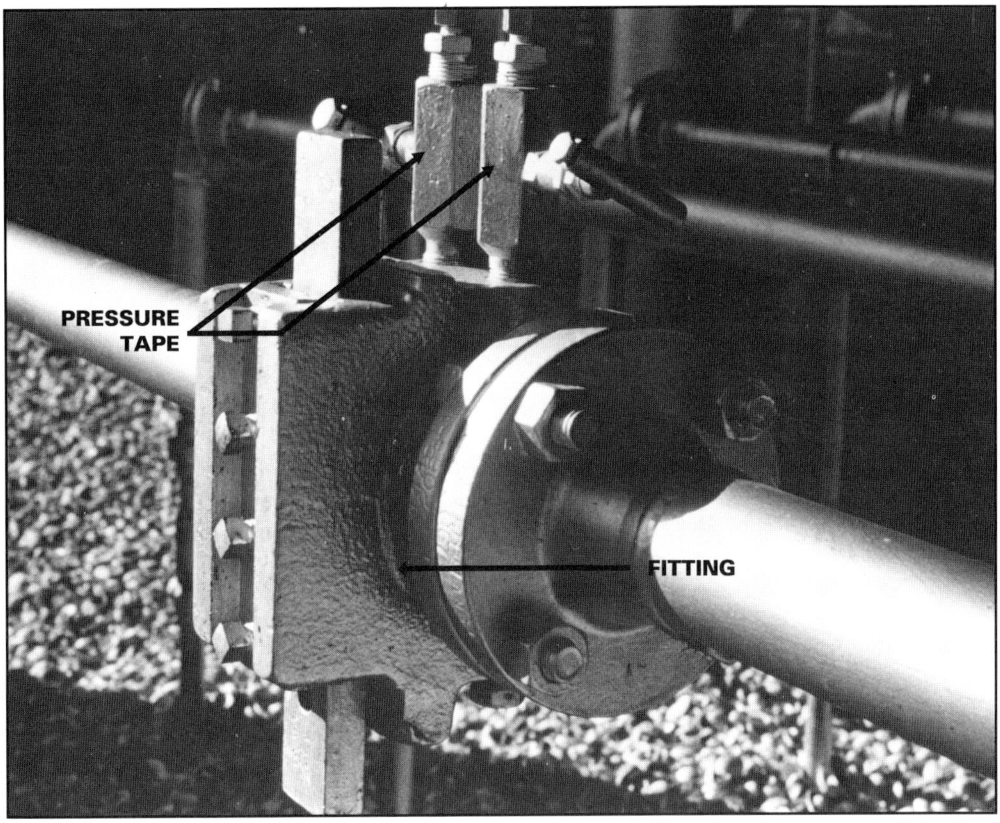

Figure 8.15. Meter tube and fitting installation

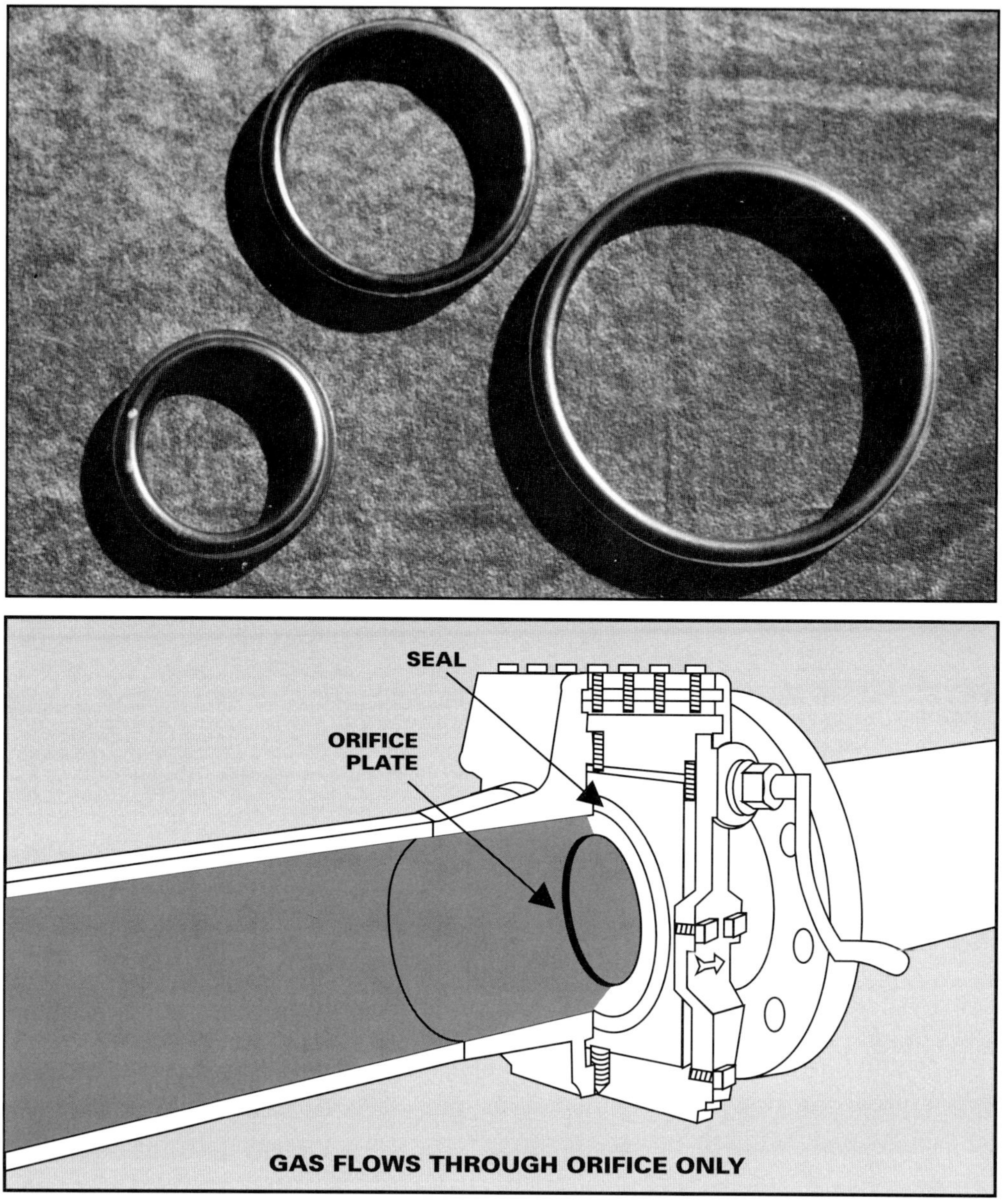

Figure 8.16. Orifice plate seals

Seals

Another important part of the primary element is the seal that keeps gas from leaking around the plate (fig. 8.16). Plate seals come in calibrated sizes and are made of different materials. In order to function as designed, the seal must be the correct size for the orifice plate and meter run, in excellent condition, and installed correctly. If any of these conditions are not met, gas may leak around the plate and produce a differential pressure that is lower than it should be and result in inaccurate calculations.

Plate and Fitting Requirements

Since the plate creates the pressure drop necessary to measure gas flow, it must meet the standard's requirements and be in excellent condition. Any damage to the plate makes gas measurement inaccurate. Even a single nick on the upstream edge of the plate can mean thousands of dollars lost through inaccurate measurement.

In order for the plate to function as designed, it must—

1. be sized correctly. If the size of the orifice does not match the size marked on the plate, gas measurement will not be accurate. The size of the orifice is a major factor in gas flow calculations. If the plate is mismarked, then you will be using an incorrect figure in your gas calculation formula. Use an inside micrometer to measure the orifice (fig. 8.17). You can tell if the plate is correct by measuring the diameter of the orifice on the upstream side of the plate. You can use a steel rule to measure the diameter of the orifice but a steel rule gives you only an approximate measure. Using a micrometer is a precise way to measure the orifice; just be careful not to nick the plate edge when using it.
2. be installed correctly. Most orifice plates have the word *INLET* stamped on the upstream side (fig. 8.18). If the plate is not marked, make sure the sharp edge faces upstream.

Figure 8.18. Upstream side of orifice plate

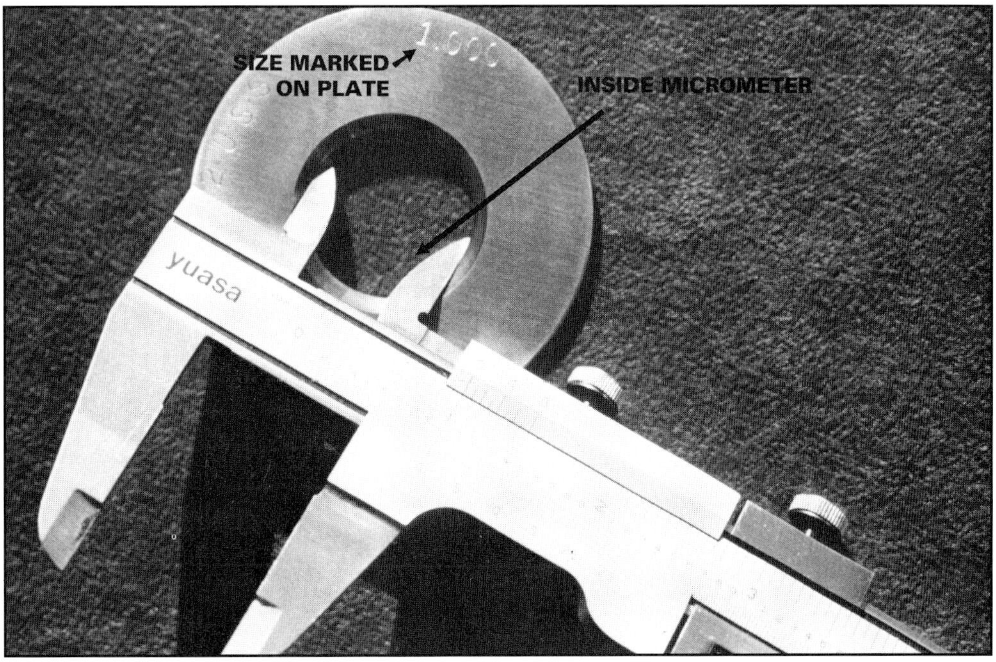

Figure 8.17. Inside micrometer

3. be as flat as possible. If the plate is warped, replace it. One good way to test a plate is to lay a steel rule across its face. If you can see light between the straight edge and the plate, the plate is warped. Try to rock the plate against the straight edge; if you can feel any rocking motion, the plate is warped. API *MPMS,* chapter 14, section 3 gives specific limits of flatness for orifice plates along with testing methods.
4. have an upstream orifice edge that is square and sharp. It should be completely free of any nicks or distortions. Run your fingernail around the edge; it should be easy to detect any major defects. Bright sunlight will reflect from the edge when the plate is held at an angle. If you see a reflected bright band, the edge is worn or damaged. A more definitive examination can be performed using a "beam of light gauge." Place a T square through the orifice and align the sharp edge of the square with the upstream edge of the orifice so that the edge of the square and the orifice are flush. Take a beam of light and shine it behind the square edge and the plate and check to see if any light comes through between the edge of the square and the upstream edge of the orifice. If it does come through, the orifice edge is not square or sharp.
5. be free of dirt, ice, or any other foreign matter, including oil. Run your finger across the face of the plate to feel for deposits and wipe clean with a soft cloth if necessary.
6. be the correct thickness. Each size orifice has a corresponding standard for thickness. A plate that is thicker or thinner than the standard's requirement may change the differential pressure reading. Use a micrometer to measure the thickness, then compare the measurement with the readings given in *AGA-3.*

If the plate does not meet all of these criteria, replace it.

Plate and seal defects result in inaccurate measurement, usually in the form of a differential that is lower than specified. The following examples are not precise but are given to illustrate a range of errors possible. A mismarked plate may cause a 50% error in the amount measured. A plate installed backwards (that is, with the sharp edge downstream instead of upstream) may produce a differential 25% lower than it should be. A warped plate may cause a 15% error. A dull or nicked plate may have a 5% error. A slightly leaking seal may cause a 2%–5% error, debris on the plate and a plate with the incorrect thickness may introduce a smaller percentage error.

Figure 8.19. Orifice plate storage box

In terms of money, errors can mean great losses. For instance, with 900,000 cubic feet per day flowing through a mismarked plate with a 50% potential error, and with gas selling at, say, $5.00/thousand cubic feet (Mcf), that's a potential dollar impact of $2,250/day (.5 × 5 × 900), or over $821,250/year. If the plate is installed backwards, the same amount of gas flowing at the same price can mean a loss of $1,125/day, or over $410,000/year. Even smaller margins of error are costly. A leaking seal can mean a lot of money lost—in our example, $90/day, or $32,850/year.

It is very important that the plate and seal be clean and in excellent condition. The plates should always be stored in a box designed especially for them (fig. 8.19). The storage box should be designed so that the plates are stored vertically and separated from each other by a nonmetallic divider, usually wood. Never hang plates on metal surfaces like manifold valves, fitting-gear shafts, or nails. Metal-to-metal contact may damage the orifice edge.

Figure 8.20. Secondary element

THE SECONDARY ELEMENT

The secondary element records the conditions existing in the primary element. Electronic computers are being used more often as the secondary element but, in many areas the secondary elements consist of sensing or gauging lines and a recorder that houses the differential pressure–sensing element, an orifice meter chart, a clock, a static spring, and a pen and ink system (figs. 8.4, 8.20).

Installation

For any secondary element to work as designed, it must be installed correctly. The element, as shown in figure 8.20, should be installed so that the pressure tap holes face upward and the gauge lines rise from the tap holes vertically at a slant of at least 1 inch per foot of gauge line, without sags or loops. This allows any liquid in the lines to drain back into the meter run. Some meter installations have drip pots to collect the liquids that do not drain back into the meter tube (fig. 8.21). The drip pots should be drained or bled regularly to prevent liquids from accumulating and being trapped in the gauge lines, thereby causing erroneous readings. If liquid is trapped in the upstream gauge line, it blocks the upstream pressure from the differential sensing element. The result is a differential pressure reading that is lower than it should be. If liquid is trapped on the downstream side, the result is a differential pressure reading that is higher than it should be.

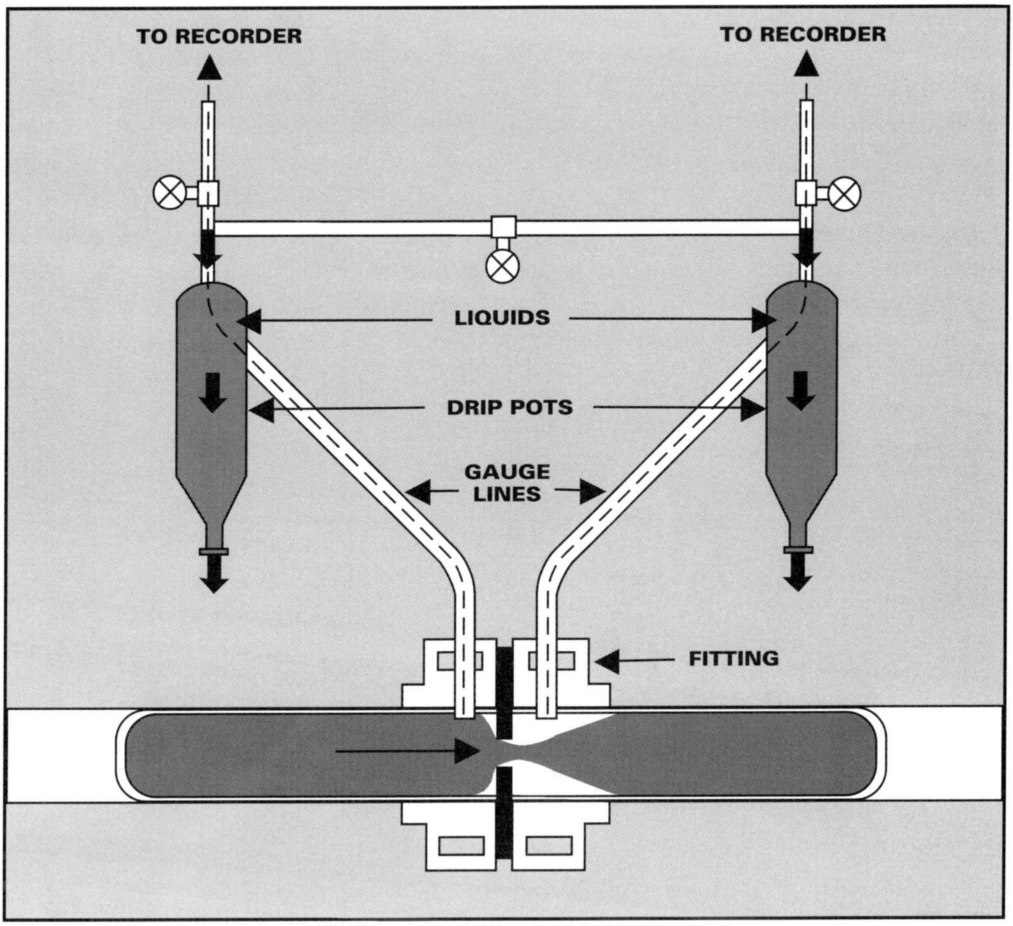

Figure 8.21. Drip pots

The recorder housing, the glass door, and the seal around the door should all be in good condition so moisture is kept out. Moisture can ruin the meter chart and cause rust and corrosion to form on the recorder clock, the static spring, and other elements that record the variables necessary for gas calculations.

Both the primary and the secondary elements of an orifice meter installation should be located in an area that is free of mechanical vibration and pulsating flow. Pulsating flow can be caused by reciprocating compressors and regulators, separator dumps, intermitting wells, and so on, and can lead to wide line charts and inaccurate measurement. If the orifice meter installation is located near this type of equipment, it may need to be relocated or the gas-flow process modified by placing as many vessels or pipe lengths between the pulsation source and the meter as possible. Vibration can also result in messy charts, but this problem can be solved by mounting the meter installation or the recorder with spring-loaded shock mounts.

Orifice Meter Charts

The meter chart (fig. 8.22) is the record of the variables needed to calculate the amount of gas flowing through the orifice plate during a given time period. These variables include the static pressure, the differential pressure, and the time of day these pressures were recorded. In some cases, the temperature of the gas may also be recorded. The differential is measured in inches of water and is recorded on the chart in red ink. The static pressure is the flowing pressure in the line measured at one of the pressure taps. It may be measured either upstsream or downstream (the most common) from the plate in psig or psia and recorded on the chart in blue or black ink. Since charts are used to calculate gas flow during a particular time period, time divisions are printed on the outside edge. Charts can measure 24 hours in a single rotation, like the one in figure 8.22, or multiple days, like the one in figure 8.23.

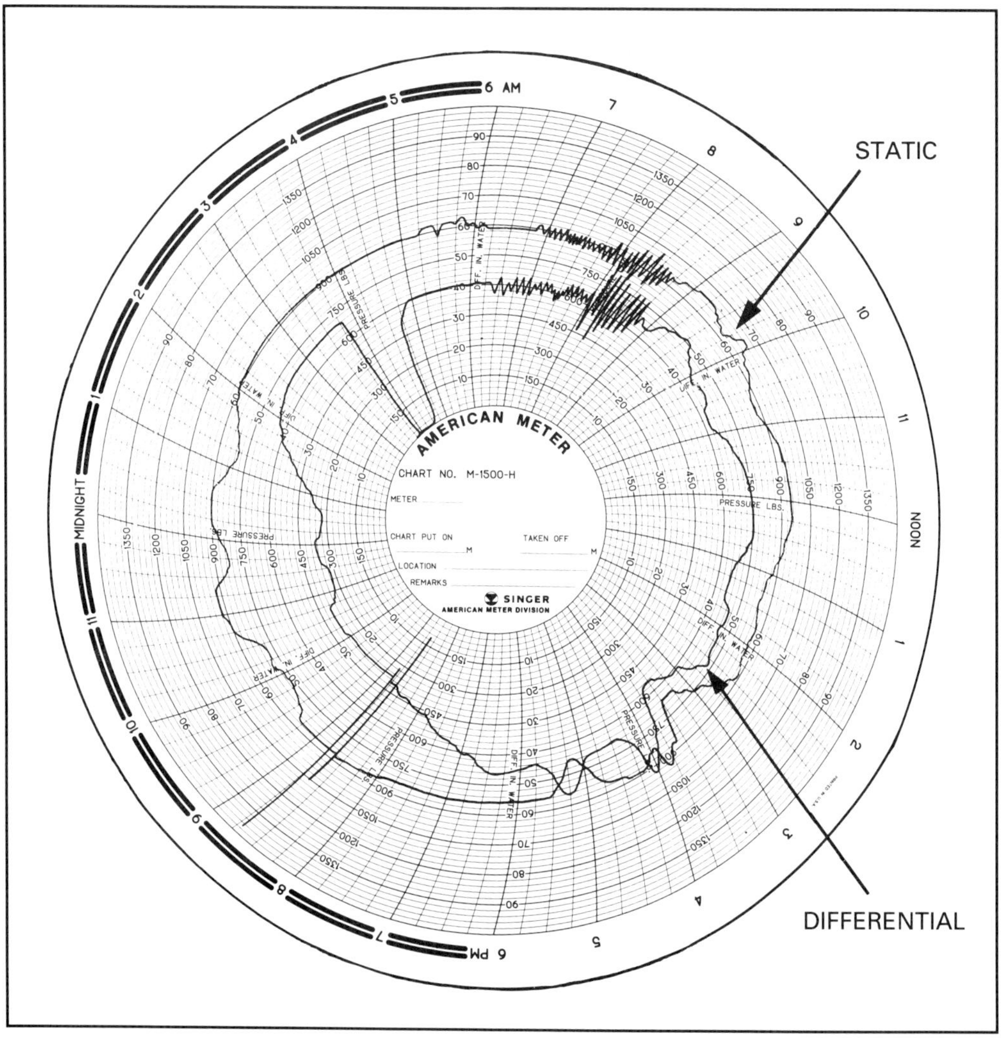

Figure 8.22. Twenty-four-hour meter chart

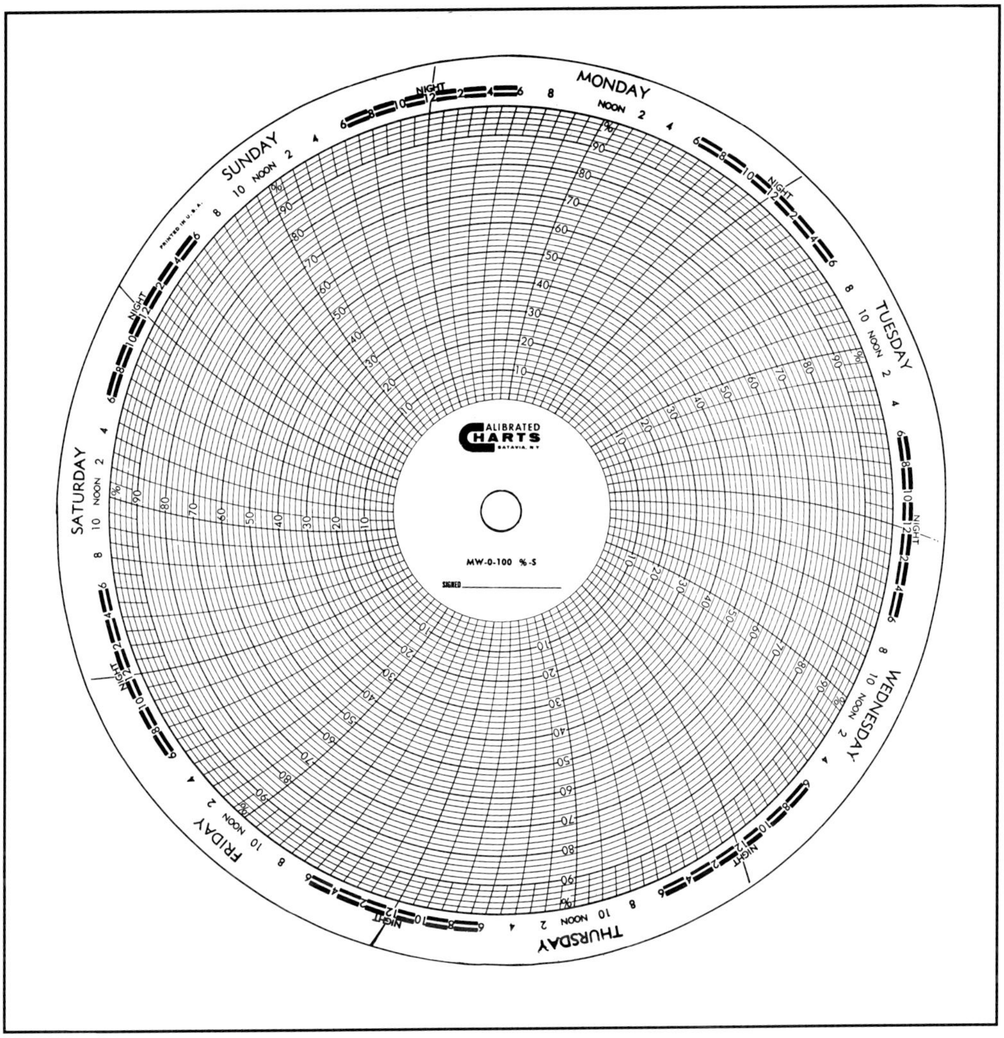

Figure 8.23. Seven-day meter chart

The gas temperature may also be recorded, since it affects the gas flow calculation. If the temperature is recorded on the same chart as the differential and the static pressures, it is recorded in green. If the temperature is recorded on a separate instrument, the color of the ink is not critical. If the temperature is not recorded on a chart, read and record the gas temperature at the time the chart is retrieved.

There are three types of orifice meter charts: the linear; the L-10, or square root; and the percentage. Each records the same variables in a different way. The linear chart (fig. 8.22) is a direct reading chart. *Direct reading* means that the static and the differential pressures can be read directly from the chart. For example, if the differential pressure is 50 inches of water, then it is recorded as 50 inches on the chart at a given point in time, for example, at 9:30 AM. And, if the static pressure is 960 psig, then it is recorded as 960 psig, for example, at 8:00 AM. This chart has uniform divisions with a scale for the differential pressure that corresponds to the differential range of the meter, and a static scale that corresponds to the static range of the meter.

Linear charts must exactly match the differential and static ranges of the flow recorder; otherwise, the readings used for volume calculations will not be accurate. For instance, if the meter's differential range is 0 to 100 inches of water, as in figure 8.22, then the chart's divisions must read from 0 to 100 as well. And, if the meter's static range is 0 to 1,500 psig, then the chart must read from 0 to 1,500 on the static scale. The differential pressure range is stamped on the recorder's identification (ID) plate (fig. 8.24). The ID plate should be updated any time a change in differential pressure range is made. The static range of the element is stamped on the ID plate of the recorder (fig. 8.25) and on the static spring (fig. 8.26). It is best to identify the static range by what is stamped on the spring. If the spring is changed, the change should be stamped on the ID plate, but sometimes this last step is inadvertently omitted.

Figure 8.24. Differential pressure range is stamped on the ID plate of the recorder.

Figure 8.25. Static pressure range is stamped on the ID plate of the recorder.

Figure 8.26. Static pressure range is stamped on static spring.

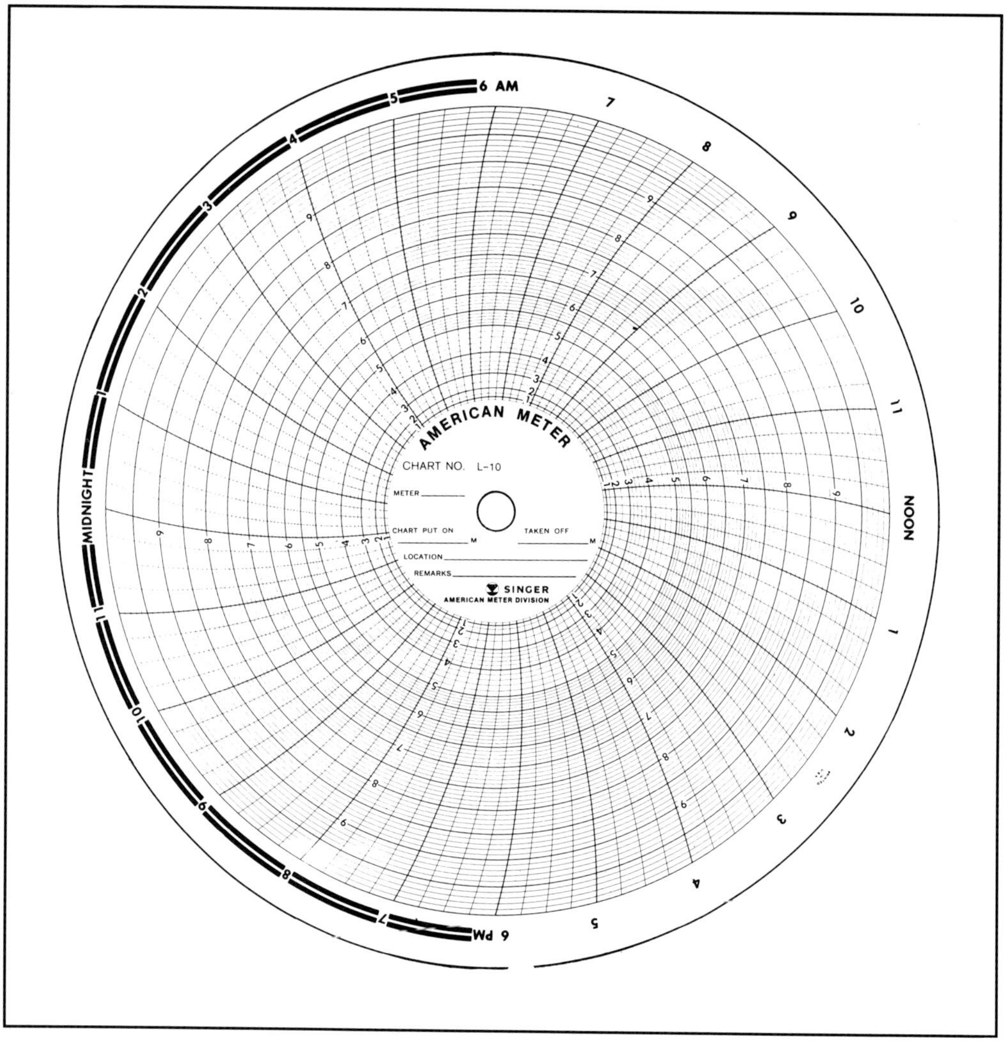

Figure 8.27. L-10 meter chart

The second type of chart is the L-10, or square root, chart (fig. 8.27). This chart is *not* a direct reading chart. Pressures are recorded on a logarithmic scale from 1 to 10. You use the same scale for both static and differential pressures, so you can use this chart on recorders of all ranges.

Since this chart is not a direct reading chart, you cannot tell what the pressures are just by reading the chart. You have to convert the readings mathematically to determine the pressures. For example, suppose you have an L-10 chart with a differential reading of 5. That 5 does not mean 5 inches of water. To convert this reading to actual differential, first you square the reading; in this case, 5 squared is 25. Next, you multiply the product by the differential range of the meter. If the range is 0 to 200 inches of water, multiply 25 by 200 to get 5,000. Then, since any reading on an L-10 chart is the square root of 100% of the range of the meter, divide by 100 to convert to differential. In this example, a reading of 5 means a differential of 50 inches of water.

To convert a static reading, you follow the same procedure. If the static reading is 8, square 8 to get 64. Multiply 64 by the static range, say, 1,500 psig, to get 96,000. Divide this product by 100 to get a static pressure of 960 psig.

The third type of chart is the percentage chart (fig. 8.23). It is not a direct reading chart, either. The readings on this chart must also be converted mathematically to determine the pressures. It has uniform divisions from 0 to 100 for both the differential and the static readings. Each line on the chart represents a percentage of the static or differential range of the meter. For example, if the chart reading indicates the differential is 40% (e.g., at 6:00 AM), then the actual differential is 40% of the meter's range. For a meter with a differential range of 0 to 200 inches of water, the differential pressure is 80 inches of water (40% of 200 is 80). A static reading of 50% means 50% of the static range of the meter. If the static range of the meter is 0 to 1,500 psig, the static pressure is 750 psig (50% of 1,500 is 750). This type of chart can also be used on meters with different ranges.

Any of these three charts can be printed with lines that are dark enough to read easily, or printed with faint lines to be read, or scanned, by an optical scanner. Charts with faint lines are called scanner charts. If the chart is a scanner chart, write only on the back of it unless specifically directed to do otherwise. The scanning system is designed to read anything written on the face of the chart, so any writing on the face interferes with the scanner's ability to integrate the information recorded on the chart.

In addition to manufacturing different types of charts, some companies produce charts with different zero points. In figure 8.28 the chart's zero line is 1⅛ inches from the center of the chart. Other manufacturers place the zero line ¾ of an inch from the center. These different starting points affect the entire scale. The appropriate chart for each meter manufacturer must be used if gas measurement is to be accurate. The pen on the recorder must be made to match the zero point and the span on the chart (see item 7 of the next section), or measurement will not be accurate.

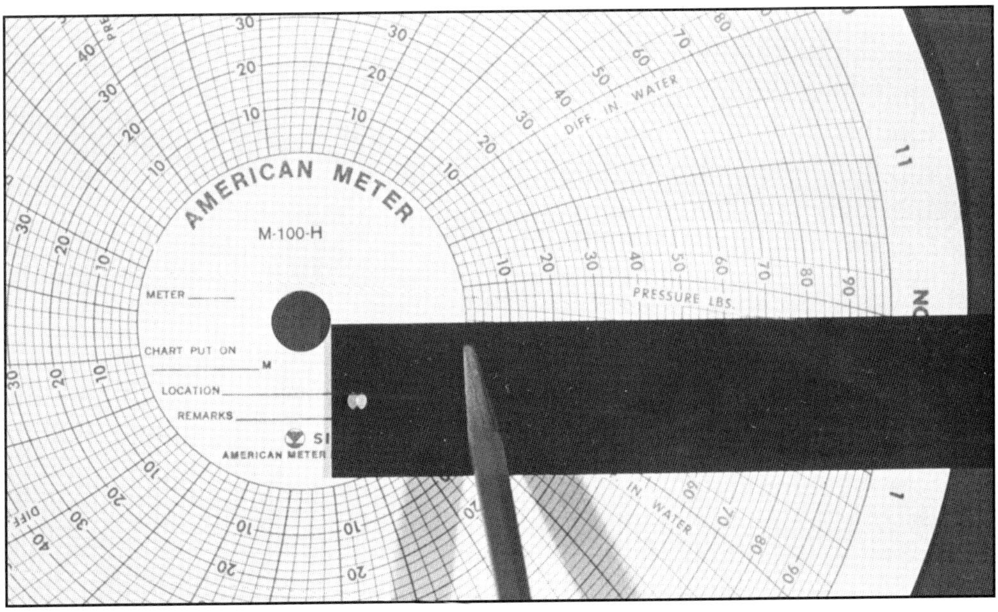

Figure 8.28. Differential zero point on chart

Producing Accurate Orifice Meter Charts

In order to make sure that the chart on the recorder has all the information needed to calculate gas flow, you must—

1. make sure the clock is accurate. When changing the chart, note the beginning and ending times and make sure the chart has rotated 360° in that time period. If it has not, gas measurement will not be accurate. For instance, if the clock is running slow, the chart will not rotate through a complete cycle in the allotted time. If the clock is fast, the chart rotates too far and the data overlap. Since gas measurement is made in cubic feet or pounds mass (depending on which formula you are using to calculate gas flow) per time unit, any clock problems will cause inaccurate calculations. Note any clock problems on the chart so the integration department will know what happened and make the proper adjustments to their calculations, then adjust or replace the clock.

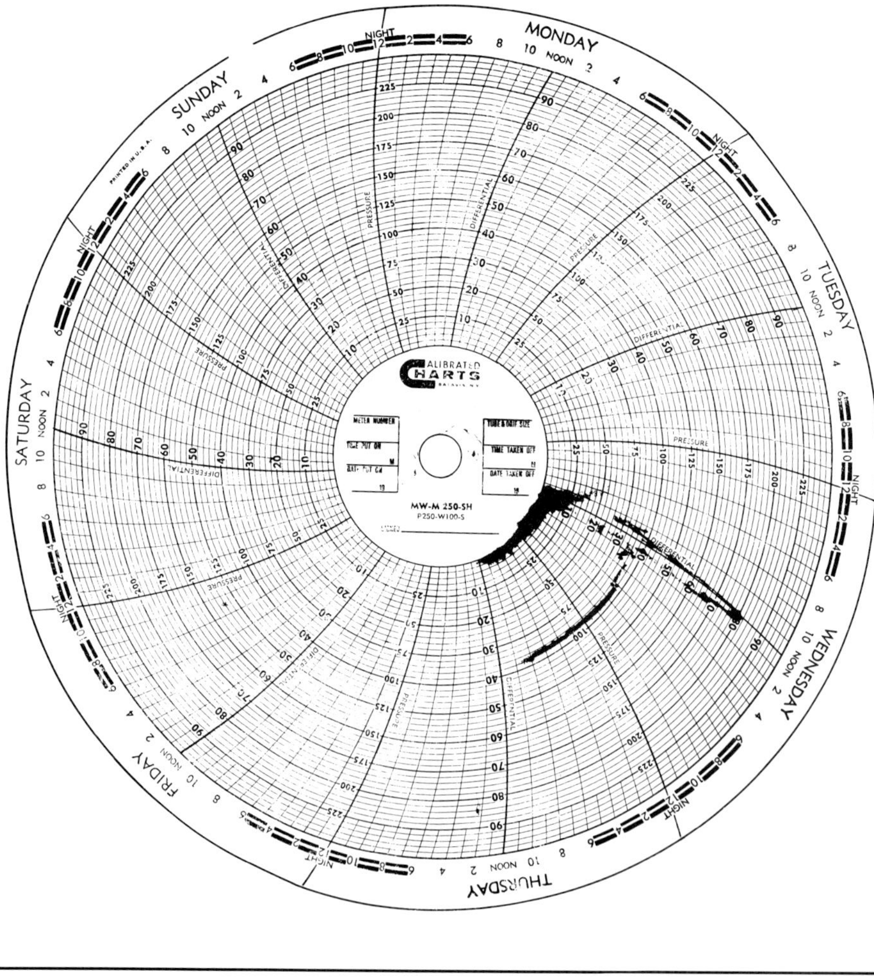

Figure 8.29. Water-damaged 7-day linear chart

2. mark the chart clearly with such identifying items as the station number, the orifice plate size, the name of the lease, time on and off, date on and off, location, static, differential, and temperature (if appropriate) ranges, and the lease operator's initials. At some meter stations, you write only the station number and time and date on and off. The rest of the data are in the meter station's records.
3. note anything unusual in the flow rate. Write these notes on the face of the chart without interfering with the recorded lines, *unless* the chart is a scanner chart, in which case, write on the back. Note any unusual occurrences, such as a stopped clock, a frozen meter, or liquid in the gauge line. You are the only one who knows what has happened, so help the chart integration department all you can.
4. take special care when handling the chart, after pulling it from the recorder and noting the necessary information. Keep it in good shape; do not fold or bend it, and do not put it inside your shirt to try to protect it. Blot the chart so the ink will not smear. Do not let it get wet, like the one shown in figure 8.29. Charts tend to absorb moisture and swell, making them inaccurate and difficult to integrate.
5. store the chart after removal from the recorder in a container made of plastic or some other durable material. The container keeps the chart clean and dry and protects it from damage.
6. check the recorder's pens before putting a new chart on the recorder. Some recorders use disposable pens that can fray or dry out. Make sure ink is flowing correctly. If it is not, change the pens. Before you do, notice that the static pen leads the differential pen by one time division, or 15 minutes on a 24-hour chart (fig. 8.30). Make sure you retain that spacing when the pens are replaced. This time lead is the same one used on the chart integrator. The lead times must match. Some recorders use capillary ink systems. The reservoir in these systems needs to be refilled from time to time. If this

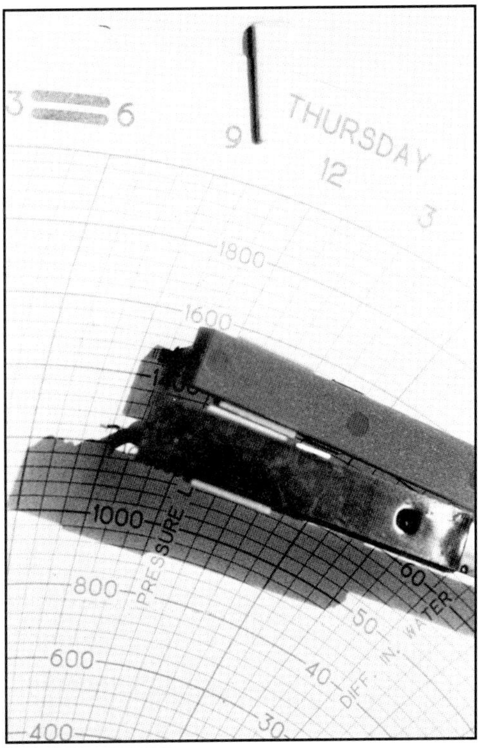

Figure 8.30 Differential pen leads static pen by one time division.

type of system is not properly maintained, you can get ink runs, which cause messy, hard-to-integrate charts or gaps in information. Keep the capillary systems serviced.

7. check the zero point for the differential pen each time you change the chart. By checking the zero point, you ensure that each recording of the differential pressure is traceable to a base point. Always zero the differential pen with line pressure on the meter. In other words, zero the differential pen at working static pressure. Never depressure, or bleed down, the meter before zeroing the differential pen. The meter should be depressured only by a qualified meter inspector or measurement technician trained in measurement procedures. To zero the differential pen, first close the bleeder valve on the meter manifold (fig. 8.31). (The bleeder valve should always be open during normal operations.) Next, open the two bypass valves. Then, close either the upstream or the downstream meter-run tap valve. Closing it should trap and equalize pressure on each side of the differential recorder and move the red differential pen down toward zero. If it does not zero perfectly, that is, line up with the zero point on the chart,

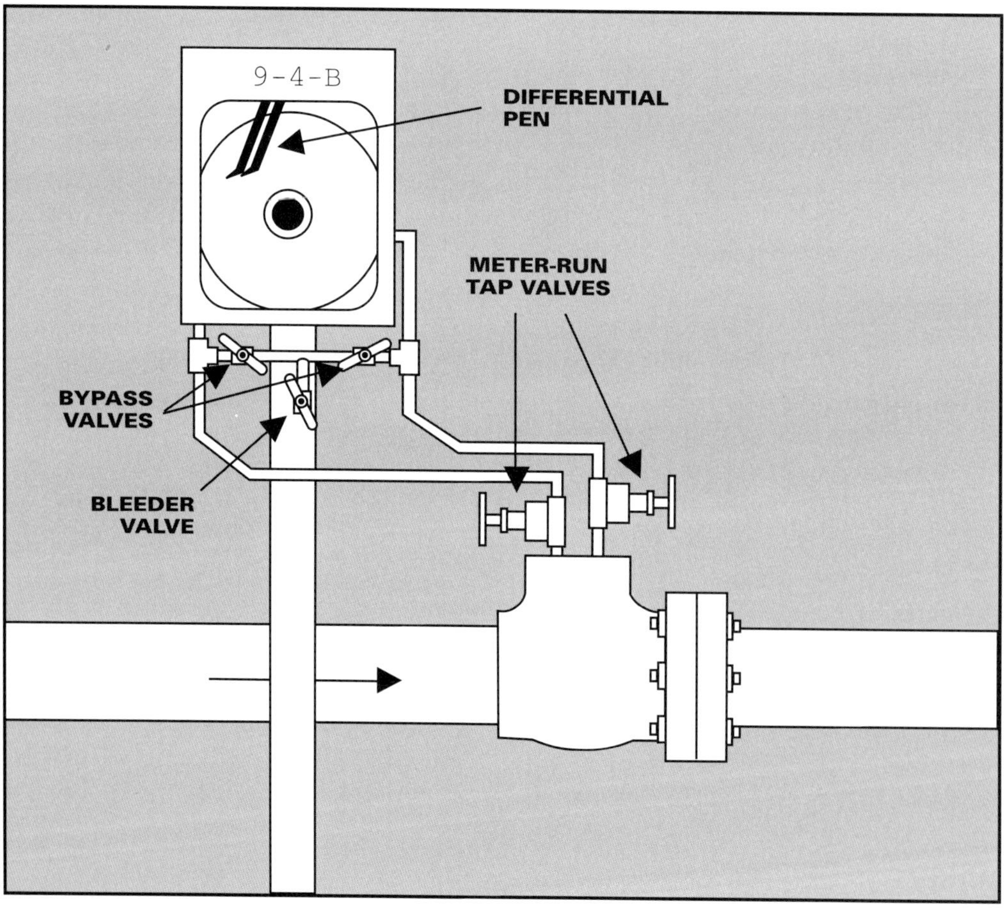

Figure 8.31. Secondary element flow recorder

close the other tap valve and check the zero again, then adjust the thumbscrew on the recorder, or take a screwdriver to adjust the pen to the zero position. If necessary, turn the micrometer screw at the top of the red pen arm, then rotate the chart counterclockwise one time division to make sure the pen is zeroed exactly.

To return the recorder to service, slowly open any meter-run tap valves that you closed before, then close the bypass valves. At this point, the differential pen should move back to the operating position—that is, recording the differential pressure. If there is a major change in differential pressure reading from the previous reading, check the cause and note it for the integration department. Finally, open the bleeder valve and check for leaks. Leaks across this valve cause incorrect differential readings. When opening or closing any valve, remember not to screw the stems down too hard against the seats to avoid damaging the valves.

Electronic Flow Measurement

As stated earlier, many locations are changing the secondary element of the orifice meter station from pens and charts to field-mounted gas-flow computers. There are several reasons for this conversion. One reason for switching to electronic measurement is that chart recorders cannot provide real time gas-flow data. Thus, you cannot tell from a remote location what the current differential pressure, static pressure, flowing temperature of the gas, or gas volume is. You must travel to the site, retrieve the chart, and return it to the integration department before the data can be processed and the amount calculated. All of these prerequisites require labor expense and a lengthy delay, which prevents timely calculation of gas flow. In addition, the last requirement—returning the chart in excellent condition—is not always easy to achieve. The charts tend to curl up when the humidity changes, the ink runs when the paper gets wet, chart motors stop because someone forgot to wind them or to replace run-down batteries, so you do not have data for some of the time the chart was on the recorder

Electronic recording of the measurement data eliminates many of these problems, but for years its implementation was hampered by a lack of power. Running electricity to many remote locations is impractical. But with the development of low-power gas-flow computers that can be run by a pair of ordinary alkaline lantern batteries that last well over a year, or by inexpensive solar panels with rechargeable batteries, electronic measurement has gained in popularity. (Battery and solar panels should be well maintained to prevent a power failure and inaccurate measurement.) The advent of 16-bit microprocessors, inexpensive memory, and "smart" differential and static pressure transducers has finally made accurate results practical.

Basic Field Computer Operations

Most field computers calculate the gas flow at short fixed intervals—normally, every second. They use the differential and static pressures and the flowing temperature of the gas measured at the orifice meter run to compute gas flow. In addition, they use a number of other factors that have been fed into the computer and stored: the orifice diameter, the pipe diameter, the specific gravity of the gas, and any other data required to calculate the gas flow.

The heart of the electronic system is the transducer. Transducers convert the differential and static pressures and the flowing temperature of the gas to electronic signals. These signals are fed into a digital computer with display (fig. 8.5). The computer makes calculations and stores the information while the display gives a readout of the stored total gas flow. Most computers have enough memory for approximately thirty-five days.

To recover the data stored in the flow computer, you can use a portable computer or handheld recording device furnished by the manufacturer that plugs into the on-site computer. Using the portable system's keyboard and display, you can retrieve the data you need.

You can print a hard copy of the data by connecting the portable computer to a printer, or you can transfer the data to a database in a larger computer. You can also use the portable unit for entering and changing data stored in the field computer. Many companies are providing field operators with laptop or desktop computers in lieu of the handheld unit shown in figure 8.5. Both the laptop and the desktop can be used with software that allows the field operator to use the unit for a variety of purposes.

Some electronic measuring systems are equipped with flow computers that you can access by telephone, radio, or satellite. In other words, you can call your on-site computer from a different location, retrieve any data you need, and make any changes in the stored data without having to travel to the site. This remote capability gives you access to real-time data from any location.

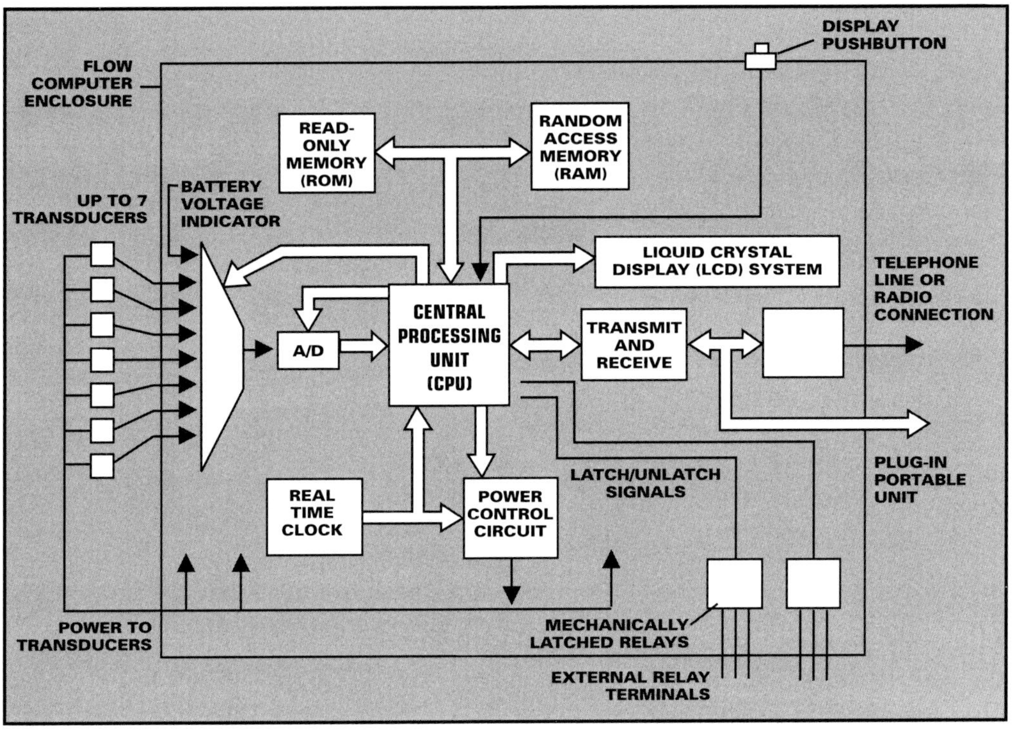

Figure 8.32. Simplified block diagram of a low-power field-level computer system

Basic Requirements of Low-Power Field-Level Computer Systems

Electronic systems can vary, but a basic system (fig. 8.32) consists of a set of transducers, connecting wires, a field computer (much like the computers found in any office, except these computers are weatherized and explosion-proof), and some type of hand-held dedicated-access device, laptop, or desktop computer. Transducers are often mounted on or near the orifice run and outside the computer housing for flexibility and ease of access. Multiple orifice meter runs use individual differential and static pressure transducers and individual or shared temperature transducers. Most field computers include mechanically latched relays. These relays provide a way to add external equipment such as gas analyzers and odorant injectors. You can also add radio or telephone data communication at any time by providing an appropriate radio modem.

CALCULATION OF GAS FLOW

Gas flow is calculated using the same formulas, regardless of the source of the variable data. If your meter station is equipped with a computer, the calculations are done on site; if your meter station is equipped with a chart recorder, the calculations are done at another location. The current API-AGA Standards (adopted in August 1992) recommend new formulas for calculating gas flow.

The formulas recommended prior to August 1992 proved satisfactory for years, given the methods used to find and to record differential and static pressures. Much time and training was devoted to teaching field operators to handle the orifice meter charts properly so that the gas volume flow could be calculated as accurately as possible. Because of the nature of gas, however, and the fact that recording on a chart can be only so precise, some equation factors that affect gas measurement only slightly were not included in field calculations. With the development of on-site computers, which make the collection of real-time data possible, all of the factors can now be included in the calculation of the gas flow. As a result, API *MPMS*, chapter 14, section 3, published in August 1992, recommends new formulas for determining gas flow. These formulas include the factors in the previously recommended formulas in addition to several others designed to improve gas measurement. The differential pressure, the static pressure, and the flowing temperature are still used, along with other factors, to calculate the amount of gas flowing through the meter. The standard now recommends that the amount of gas flowing through the meter be calculated directly either with a formula measuring the mass flow rate or with a formula measuring the volume flow rate. The new formula for measuring mass flow rate (the practical mass equation) follows:

$$Q_m = N_1 C_d E_v Y d^2 \sqrt{\rho_{t,p} \Delta P}$$

where—

Q_m	=	mass flow rate
N_1	=	unit conversion factor, the value of which depends on the mass units chosen
C_d	=	orifice plate coefficient of discharge
E_v	=	velocity of approach factor
Y	=	expansion factor
d	=	orifice plate bore diameter calculated at flowing temperature (T_f)
$\rho_{t,p}$	=	density of the fluid at flowing conditions ($P_f T_f$)
ΔP	=	orifice differential pressure.

Each of these factors accounts for certain characteristics of gas as it flows through an orifice meter as well as characteristics particular to an installation, such as orifice plate size and meter run size. To determine numerical values for each of these factors, you must consult the *API Manual of Petroleum Measurement Standards,* chapter 14, section 3, "Natural Gas Fluids Measurement, August 1992," or *AGA-3*. In these manuals you will find formulas to determine each of the factors in the practical mass equation. If you are using an electronic measuring system, all of the numbers are collected and stored in the computer, which calculates the mass flow rate for you. You can access the computer to determine the mass flow rate or to determine any of the individual factors used to calculate the mass flow rate. Obviously, calculating the gas flow with the new formulas is easier if left to the computer, but, if required, you can use the various formulas given in the API manual and calculate the flow yourself.

Determining gas flow with a mass formula results in a direct measure of the gas independent of a standard pressure and temperature. The answer to a mass flow rate calculation is not given in cubic feet per time unit at a specified pressure and temperature, as the volume measurement is. Instead, the mass flow rate is expressed in *pounds mass per unit time.*

Once you determine the mass flow rate, you can convert the measure to a volume flow rate, if necessary, by using the following formula:

$$Q_v = Q_m \div \rho_b$$

where—

Q_v = volume flow rate

Q_m = mass flow rate

ρ_b = density of fluid at base conditions.

The volume flow rate may be directly calculated from equations in API *MPMS*, chapter 14, section 3 and may be used for accounting consistency. The volume flow rate formula is discussed later in this chapter. The results with either method yield equal results.

Mass Flow Rate and Volume Flow Rate

The new formulas adopted in 1992 allow you to calculate a *mass* flow rate or a *volume* flow rate. Both rates are different measures of the same amount of material.

To explain, suppose you have a large number of bricks to ship and you want to determine exactly how many. Also suppose your company has 10 trucks and each truck holds 10,000 bricks. You could determine the number of bricks by filling each truck and counting the number of trucks. If you filled 10 trucks, you would have 100,000 bricks.

But what if you were measuring this same amount of bricks, but each truck held only 8,000? Ten trucks would mean 80,000 bricks. In order for a buyer and a seller to determine the total number of bricks, they would have to agree on how much each truck held. This type of measurement would be a volume measurement based on a standard number of bricks per truck.

The buyer and the seller could also measure the amount of bricks a different way. Suppose each brick possesses a known measurable characteristic—say, it emits a specific amount of radioactivity. You could measure the number of bricks by measuring the total amount of radioactivity emitted by the bricks shipped. Measuring the radioactive output is a different, but just as accurate, method of counting the same bricks.

Natural gas is not radioactive, of course, but the same principle applies to measuring the quantity of gas flowing past the orifice plate. You can measure this amount in different ways.

When you measure the *volume* of gas flowing past an orifice meter, you are measuring the number of cubic feet of gas that have passed the orifice plate in a certain amount of time. The question is, how many gas molecules are in each cubic foot? Remember, the answer depends on the temperature and the pressure of the gas. In most sales contracts, the buyer and the seller must agree on a standard pressure and temperature before any exchange takes place. In most areas it is 14.73 psia at 60°F, but in Louisiana it is 15.025 psia at 60°F. In other areas, it is 14.65 psia at 60°F. In fact, it could be any pressure or temperature the buyer and the seller agree on. In the formulas recommended before 1992, one of the factors used was a pressure base factor. This factor is a number used to adjust measurement to a standard presssure agreed on in the sales contract. In the new formulas for determining gas flow, a pressure base factor is assumed constant at 14.73 psia when you are calculating the *volume* that has passed the orifice plate.

The volume flow rate equation in API *MPMS,* chapter 14, section 3, August 1992, is as follows:

$$Q_v = 7709.61 C_d(FT) E_v Y_1 d^2 \sqrt{\frac{P_{f1} Z_s h_w}{G_r Z_{f1} T_f}}$$

where—

Q_v = volume at standard conditions of 14.73 psia and 519.67°R (Rankine), in cubic feet/hour.

7709.61 = numeric constant

$C_d(FT)$ = coefficient of discharge for flange-tapped orifice meter

E_v = velocity of approach factor

Y_1 = expansion factor (upstream tap)

d = orifice bore diameter, in inches, calculated at flowing temperature (T_f)

P_{f1} = flowing stream pressure (upstream tap), in psia

Z_s = compressibility at standard conditions ($P_s T_s$)

h_w = orifice differential pressure, in inches of water at 60°F

G_r = real gas relative density (specific gravity)

Z_{f1} = compressibility at upstream flowing conditions ($P_{f1} T_f$)

T_f = flow temperature, in ° Rankine.

This calculation assumes that the base and standard conditions are 14.73 psia and 519.67°R. If, however, base conditions are different from standard conditions, the volume flow rate calculated at standard conditions can be converted to the volume flow rate at base conditions with the following :

$$Q_b = Q_v \left(\frac{P_s}{P_b}\right)\left(\frac{T_b}{T_s}\right)\left(\frac{Z_b}{Z_s}\right)$$

where—

Q_b = base volume flow rate, in cubic feet/hour

Q_v = standard volume flow rate, in cubic feet/hour

P_s = standard pressure, in pounds force psia

P_b = base pressure, in pounds force psia

T_b = base temperature, in °Rankine

T_s = standard temperature, in °Rankine

Z_b = compressibility at base conditions (P_b, T_b)

Z_s = compressibility at standard conditions (P_s, T_s).

CONCLUSION

Measuring natural gas is much more complicated than measuring petroleum liquids. Because natural gas is invisible, odorless, and has a very low density, measurement seems impossible. Instead of measuring gas flow directly, then, you must use devices that measure certain qualities of the gas, like its pressure and temperature, and then use these measurements along with a formula to calculate the total gas flow. Calculating gas flow this way may be indirect, but if you understand how the gas-measuring equipment works, if you maintain that equipment, and if you are trained in measurement procedures, your measurements will be accurate.

APPENDIX
Pre-1992 Gas-Flow Formulas

All of the gas-flow formulas recommended by API as of August 1992 use the recorded differential and static pressure readings and the flowing temperature of the gas from the orifice meter run and combine these with several other numerical factors to calculate gas volume flow. Depending on contractual agreements, however, the formulas recommended by API before 1992 may have some field applications and are therefore included in this Appendix.

FIELD FORMULAS USING C′

In spite of the recent changes in the API *MPMS*, and the *AGA-3*, some companies may use a *simplified* version of the formula recommended prior to 1992 to calculate gas flow in the field. The simplified formula for determining gas volume flow for linear charts is as follows:

Gas volume flow for linear charts = C′ × pressure extension.

The *orifice flow constant*, or *C′*, is a numerical value that accounts for various differences between ideal conditions and the variables found in the field. This constant is determined by multiplying several factors:

$$C' = F_b F_{pb} F_g F_{tf} F_{pv}$$

where—

F_b = basic orifice factor
F_{pb} = pressure base factor
F_g = specific gravity factor
F_{tf} = flowing temperature factor
F_{pv} = supercompressibility factor.

This formula given for C′ does not include several factors that were recommended by the API *MPMS* prior to1992 because these factors were not commonly used by field-level workers. For billing purposes, however, the complete recommended formula must be used. The factors listed above include numerical values that account for variables found in the field, including changes in the relative density of the gas (F_g), the size of the orifice in the meter run (F_b), the pressure standard agreed on in the contract (F_{pb}), the flowing temperature of the gas (F_{tf}), and the supercompressibility of the gas (F_{pv}).[1] API *MPMS,* chapter 14, section 3 provides tables for each of these factors based on a 0.600 relative density gas. For instance, assume the flowing temperature of the gas is 80°F. Find 80°F in API *MPMS,* chapter 14, section 3 and then locate the numerical factor to be used in your calculations, in this case, 0.9813 (table A-1). API *MPMS,* chapter 14, section 3 has numerical values for each variable. You find the number in the report for each factor and then use the figures in the formula for finding C'.

Once you have determined C', you are ready to calculate the pressure extension. The pressure extension is the heart of the equation because, whereas C' makes adjustments for contract specifications, meter changes, and minor field conditions, the pressure extension is the major variable in determining gas volume. The pressure extension can be determined different ways, depending on the policies of the company. The following is one commonly accepted method:

1. Read the static and differential pressures recorded for each time period.
2. Convert each static pressure reading to absolute pressure, if necessary.[2]
3. Multiply the differential for each time period by the absolute static for the same time period.
4. Take the square root of that product.
5. Average the square roots of the products calculated for each time for the total chart period.

Multiply the pressure extension by the orifice flow constant for the time period and you have calculated the gas flow for a particular set of conditions on C' and for a particular time period. It could be the gas flow per hour, per quarter hour, per half hour—whatever the sales contract specifies. To determine the total amount of gas that has flowed past the meter in a certain time span, say, a day, you multiply the gas flow by the number of time divisions in that particular time period. For example, if the total gas flow is 10,254 cubic feet/hour, the total volume for a day would be 10,254 × 24, or 246,960 cubic feet/day. If you wanted to convert this total to thousand cubic feet (Mcf)/day (a standard method of reporting gas flow), you would divide the result by 1,000 to get 246.1 Mcf/day.

---

[1] C' is not a constant in that some of the factors used to derive it vary with changes in gas composition or flow. But these variations are usually small and may be considered constant in some field-level nonbilling measurements.

[2] Pressure recorded on linear charts may be recorded as gauge pressure (psig), or absolute pressure (psia). *Gauge pressure* is the difference between absolute pressure and atmospheric pressure. *Absolute pressure* (psia) is the pressure inside the pipe plus the atmospheric pressure. To convert psig to psia, simply add the atmospheric pressure to the gauge pressure. Calculations of gas volumes are based on absolute pressure.

TABLE A-1

F_{tf} Factors Used to Change from a Flowing Temperature of 60°F to Actual Flowing Temperature (excerpt from API, *MPMS*, chap. 14, section 3—August 1992)

$$F_{tf} = \sqrt{\frac{60 + 459.67}{T_f + 459.67}}$$

Temperature °F	F_{tf}	Temperature °F	F_{tf}	Temperature °F	F_{tf}
–20	1.0872	40	1.0198	100	0.9636
–19	1.0859	41	1.0188	101	0.9627
–18	1.0847	42	1.0178	102	0.9619
–17	1.0835	43	1.0168	103	0.9610
–16	1.0823	44	1.0158	104	0.9602
–15	1.0810	45	1.0148	105	0.9593
–14	1.0798	46	1.0137	106	0.9585
–13	1.0786	47	1.0127	107	0.9576
–12	1.0774	48	1.0117	108	0.9568
–11	1.0762	49	1.0108	109	0.9559
–10	1.0750	50	1.0098	110	0.9551
–9	1.0738	51	1.0088	111	0.9543
–8	1.0726	52	1.0078	112	0.9534
–7	1.0715	53	1.0068	113	0.9526
–6	1.0703	54	1.0058	114	0.9518
–5	1.0691	55	1.0048	115	0.9509
–4	1.0679	56	1.0039	116	0.9501
–3	1.0668	57	1.0029	117	0.9493
–2	1.0656	58	1.0019	118	0.9485
–1	1.0644	59	1.0010	119	0.9477
0	1.0633	60	1.0000	120	0.9468
1	1.0621	61	0.9990	121	0.9460
2	1.0610	62	0.9981	122	0.9452
3	1.0598	63	0.9971	123	0.9444
4	1.0587	64	0.9962	124	0.9436
5	1.0575	65	0.9952	125	0.9428
6	1.0564	66	0.9943	126	0.9420
7	1.0553	67	0.9933	127	0.9412
8	1.0541	68	0.9924	128	0.9404
9	1.0530	69	0.9915	129	0.9396
10	1.0519	70	0.9905	130	0.9388
11	1.0508	71	0.9896	131	0.9380
12	1.0497	72	0.9887	132	0.9372

TABLE A-1—*continued*

Temperature °F	F_{tf}	Temperature °F	F_{tf}	Temperature °F	F_{tf}
13	1.0485	73	0.9877	133	0.9364
14	1.0474	74	0.9868	134	0.9356
15	1.0463	75	0.9859	135	0.9348
16	1.0452	76	0.9850	136	0.9340
17	1.0441	77	0.9840	137	0.9332
18	1.0430	78	0.9831	138	0.9325
19	1.0419	79	0.9822	139	0.9317
20	1.0409	80	0.9813	140	0.9309
21	1.0398	81	0.9804	141	0.9301
22	1.0387	82	0.9795	142	0.9294
23	1.0376	83	0.9786	143	0.9286
24	1.0365	84	0.9777	144	0.9278
25	1.0355	85	0.9768	145	0.9271
26	1.0344	86	0.9759	146	0.9263
27	1.0333	87	0.9750	147	0.9255
28	1.0323	88	0.9741	148	0.9248
29	1.0312	89	0.9732	149	0.9240
30	1.0302	90	0.9723	150	0.9232
31	1.0291	91	0.9714		
32	1.0281	92	0.9706		
33	1.0270	93	0.9697		
34	1.0260	94	0.9688		
35	1.0250	95	0.9679		
36	1.0239	96	0.9671		
37	1.0229	97	0.9662		
38	1.0219	98	0.9653		
39	1.0208	99	0.9645		

GAS FLOW FORMULA FOR USE WITH L-10 CHART

Prior to August 1992, API recommended the following formula to calculate gas flow when pressures were recorded on an *L-10 chart*.

$$\text{Gas volume flow} = C' \times \text{a chart factor} \times \text{the average differential reading for L-10 charts} \times \text{the average static reading}$$

The last three factors make up the pressure extension for this type of chart.

C' is calculated in the same way as for a linear chart. Once you calculate C', determine the *chart factor*, a numerical value that accounts for the differences in static and differential ranges among meters. To determine the chart factor, find the absolute static range and the absolute differential range of the meter.

As mentioned earlier, the differential range is stamped on the differential spring—0 to 100 in. of water, for example. To find the static pressure range, check the static spring. It could be a number of ranges, but for this example, say it is 0 to 1,000 psia. Several publications have calculated the chart factors for a number of meter ranges; the *E-2 Handbook* is commonly used. Suppose the differential range of the meter is 200 inches of water. Locate 200 in the column marked differential range (table A-2). Locate the column for the static pressure range of the meter, 250 psi, for example. Read down the column until you come to the number across from the differential range. That number is the chart factor for this particular meter, in this case, 2.236.

Next, take the differential pressure readings for several time periods, say, each hour, add them, and then divide by the number of time periods. Then take the static pressure readings for the same time periods, add them, and divide by the number of time periods. You do not have to convert the pressure readings to actual pressures.

Finally, multiply C' by the chart factor by the average differential reading by the average static reading to determine the gas flow per time period. Again, to calculate the total flow for a particular time period, multiply the gas flow per time period by the number of time periods. To convert to Mcf, divide by 1,000.

TABLE A-2
Meter Factors, L-10 Charts

$$M = 0.01\sqrt{R_h R_p}$$

Differential Range of Meter (in.)	Maximum Range of Static Pressure Spring for Gas, Air, or Steam at Varying Pressures (psi)											
	24.7	50	100*	250	500	1,000	1,500	2,000	2,500	3,000	4,000	5,000
2.5	0.0786	—	0.1581	—	—	—	—	—	—	—	—	—
10	0.1572	0.2236	0.3162	—	—	—	—	—	—	—	—	—
20	0.2223	0.3162	0.4472	0.7071	1.000	—	—	—	—	—	—	—
50	0.3514	0.5000	0.7071	1.118	1.581	2.236	2.739	3.162	—	—	—	—
100	—	0.7071	1.000	1.581	2.236	3.162	3.873	4.472	5.000	5.477	6.325	7.071
200	—	—	1.414	2.236	3.162	4.472	5.477	6.325	7.071	7.746	8.944	10.00

Source: American Meter Company, *Orifice Meter Constants, Handbook E-2.*
*Values in this column may be used as meter factors with single-pen instruments for the measurement of liquids, or in the measurment of gas or steam at constant pressure, where the constant pressure factor, $\sqrt{P_f}$, is included as a multiplier with the orifice constant.

GAS FLOW FORMULA FOR USE WITH PERCENTAGE CHART

Prior to 1992, to calculate gas flow using a percentage chart, you first converted the pressure readings to actual static and differential pressures and then used the same formula used with linear charts. As stated in chapter 8, a reading on a percentage chart represents a percentage of the range of the meter. If the differential range of the meter is 200 inches of water and the differential reading for a particular time period is 40%, then the actual differential is 40% of 200, or 80 inches of water. The same holds true for static readings. The static reading found on the chart is a percentage of the range of the meter. Convert the reading to actual pressure and then use the actual pressure with the following formula:

$$\text{Gas flow volume for percentage charts} = C' \times \text{pressure extension}$$

Glossary

all-level sample *n:* tank sampling, a sample that is obtained by submerging a stoppered beaker or bottle to a point as near as possible to the draw-off level, then opening the sampler and raising it at a rate that makes it about three-quarters full (maximum 85 percent) as it emerges from the liquid.

allowable *n:* the amount of oil or gas that can be produced legally from a well. In a state using proration, this figure is established monthly by its conservation agency.

American Society for Testing and Materials (ASTM) *n:* organization that sets guidelines for the testing and use of equipment and materials. Its publications are *ASTM Standardization News, Book of Standards, Cement and Concrete Aggregates Journal, Composites Technology & Research, Geotechnical Testing Journal, Journal of Forensic Science,* and *Journal of Testing and Evaluation.* Address: 1916 Race St.; Philadelphia, PA 19103; (215) 299-5400.

analog data *n:* information indicated by a continuous form, usually a needle or pointer moving across a dial face. Compare *digital readout.*

angle-stem assembly *n:* a type of thermometer used on oil tanks. The graduated part of the thermometer is angled at least 90° from the temperature-sensitive portion. The angle conforms to the shell of the tank and permits easy reading of the instrument.

annealed glass *n:* glass that has been treated with a heating and cooling process that toughens it and reduces its brittleness.

API gravity *n:* measure of the density or gravity of liquid petroleum products in the United States; derived from relative density in accordance with the following equation:

$$\text{API gravity at } 60°\text{F} = [141.5 \div \text{relative density } 60/60°\text{F}] - 131.5$$

API gravity is expressed in degrees, 10° API being equivalent to 1.0, the specific gravity of water.

API *MPMS* *abbr:* American Petroleum Institute's *Manual of Petroleum Measurement Standards.*

as-built prints *n pl:* tables that provide the tank-table department with ring thickness, bottom, and appurtenance details.

ASTM *abbr:* American Society for Testing and Materials.

back-pressure valve *n:* 1. a valve used to regulate back-pressure on equipment or systems through which a fluid flows. 2. a valve used to regulate automatically a uniform pressure on the inlet side of the valve.

bad oil *n:* oil that contains a sufficient amount of sediment and water to make it unacceptable to a pipeline or other oil transportation system.

baffle *n:* a partial restriction, generally a plate, placed to change the direction, guide the flow, or promote mixing within a tank or vessel.

blade *n:* the broad flat arm of a fan, propeller, or turbine. In a displacement meter, the broad flat arm of the meter that meets with the housing to divide the liquid stream into discrete segments.

bleed *v:* to drain off liquid or gas, generally slowly, through a valve called a bleeder. To bleed down, or bleed off, means to release pressure slowly from a well or from pressurized equipment.

bleed valve *n:* a special valve used to bleed pressure slowly from a line or vessel. Also called a blow-down valve.

blending stock *n:* a petroleum product that is mixed, or blended, with another petroleum product.

block-and-bleed valve *n:* a high-integrity valve with double seals and with provision for determining whether either seal leaks.

blow-down valve *n:* see *bleed valve.*

bob *n:* see *gauger's bob.*

bottom angle *n:* the angle formed where the bottom and the side of the oil tank meet, usually nearly 90°.

bottom sample *n:* in tank sampling, a sample obtained from the material near the bottom surface of the tank, container, or line at a low point.

builder's monogram *n:* a plate on a tank built to API specifications that records the name of the manufacturer, erection date of the tank, and nominal diameter, height, and capacity.

bypass line *n:* a pipe that allows liquid to flow around a control valve, a pump, a piece of equipment, or a system.

bypass valve *n:* a valve that permits flow around a control valve, a piece of equipment, or a system.

calibration *n:* the adjustment or standardizing of a measuring instrument or of a standard capacity measure, a tank prover, or a pipe prover. Log calibration is based on the use of a permanent calibration facility of the American Petroleum Institute at the University of Houston to establish standard units for nuclear logs.

capacitance (C) *n:* the ratio of an impressed electrical charge on a conductor to the change in electrical potential; measured in farads. See *capacitor.*

capacitor *n:* an electrical device that, when wired in the line of an electrical circuit, stores a charge of electricity and returns the charge to the line when certain electrical conditions occur. Also called a condenser.

capacity table *n:* a table that tells the quantity of liquid contained in a tank at any given level.

carrier *n:* a pipeline, trucking company, or marine transportation company that transports oil or gas.

centrifugal force *n:* the force that tends to pull all matter from the center of a rotating mass.

centrifugal pump *n:* a pump with an impeller or rotor, an impeller shaft, and a casing, which discharges fluid by centrifugal force.

centrifuge *n:* a machine that uses centrifugal force to separate substances of varying densities. A centrifuge is capable of spinning substances at high speeds to obtain high centrifugal forces. Also called the shake-out or grind-out machine.

chart integration department *n:* part of a company that interprets and processes the information obtained from the orifice meter charts. The department's responsibilities include calculating the gas flow from the information on the chart.

check valve *n:* a valve that permits flow in one direction only. If the gas or liquid starts to reverse, the valve automatically closes, preventing reverse movement. Commonly referred to as a one-way valve.

closing gauge *n:* the measurement of liquid in a tank after a delivery or receipt. Compare *opening gauge.*

coefficient of discharge *n:* the ratio of actual flow to theoretical flow.

composite spot sample *n:* in tank sampling, a blend of spot samples mixed in equal proportions for testing.

condensate *n:* a light hydrocarbon liquid obtained by condensation of hydrocarbon vapors. It consists of varying proportions of butane, propane, pentane, and heavier fractions, with little or no methane or ethane.

controller *n:* an electric device used for governing the power that goes to an apparatus to which it is connected.

core sample *n:* 1. a spot sample of the contents of an oil or oil product storage tank usually obtained with a thief, or core sampler, at a given height in the tank. 2. a small portion of a formation obtained by using a core bit in an existing wellbore.

core thief *n:* see *thief.*

C_{pl} *abbr:* the correction for pressure on liquid; a factor used during a proving run to adjust for pressure on the liquid in the prover and pipe.

C_{ps} *abbr:* the correction for pressure on steel; a factor used during a proving run to adjust for pressure on the steel of the prover and pipe.

critical zone *n:* in tanks with floating roofs, the vertical range in which the gauging measurements are not accurate.

C_{tl} *abbr:* the correction for temperature of liquid; a factor used during proving to convert the temperature of a liquid to a standard temperature.

C_{ts} *abbr:* the correction for temperature of steel; a factor used during proving to adjust the temperature of the prover steel to a standard temperature.

custodian *n:* see *lease operator.*

custody transfer *n:* the changing of the ownership of or the responsibility for quantities of gas, petroleum, or petroleum products.

datum plate *n:* a metal plate attached to the tank shell on or near the bottom and located directly under the dipping reference point to provide a fixed contact surface from which liquid depth measurement can be made.

deaerator *n:* a device used for removing air or other noncondensable gases from a process stream or steam condensate and boiler feed water.

delivery tube *n:* a small device used inside a bottle when tap sampling. The tube allows for the delivery of liquid from the tank into the bottle with little or no splashing.

demulsifier *n:* a chemical with properties that cause the water droplets in a water-in-oil emulsion to merge and settle out of the oil, or oil droplets in an oil-in-water emulsion to coalesce, when the chemical is added to the emulsion. Also called emulsion breaker.

demulsifier-solvent stock *n:* a combination of a demulsifier and solvent that is used during testing for the suspended S&W content of oil and during other operations.

density *n:* the mass or weight of a substance per unit volume. For instance, the density of a drilling mud may be 10 ppg, 74.8 lb/ft^3, or 1,198.2 kg/m^3. Specific gravity, relative density, and API gravity are other units of density.

detector switch *n:* in a pipe prover, an electrical switch activated by the displacement sphere as it moves through the U-portion of the prover. Activating the switch sends an electric current through a cable to a pulse generator.

dial thermometer *n:* a thermometer where the temperature is indicated by a moving needle or pointer on a circular face or disk rather than with liquid in a glass or a liquid crystal display (LCD).

dial-type assembly *n:* see *dial thermometer.*

dielectric constant *n:* the value of dielectricity assigned to a substance. A substance that is a good insulator has a high dielectric constant, whereas a poor insulator has a low one. The dielectric constant of oil is lower than that of water, and on this principle a net-oil computer operates.

diesel fuel *n:* a light hydrocarbon mixture for diesel engines, similar to furnace fuel oil; it has a boiling range just above that of kerosene.

differential pressure *n:* the difference between two fluid pressures; for example, the difference between the pressure on the upstream side of the orifice plate and the pressure on the downstream side of the plate. Also called pressure differential.

differential range *n:* the extent of variation between the lowest and highest differential pressures that an orifice meter can accurately measure.

digital readout *n:* a type of register on which the information is indicated by directly readable characters, particularly numerals. Compare *analog data.*

direct reading chart *n:* a type of orifice meter chart that indicates the differential pressure and static pressure for a particular time period. The pressures can be read directly from the chart instead of being calculated from information given on the chart.

displacement meter *n:* a meter in which the measuring element measures a volume of liquid by mechanically separating the liquid into discrete quantities of fixed volume and counting the quantities in volume units.

diverter valve *n:* a type of valve that changes the direction of fluid flow from the intended path to an alternate route.

draw-off level *n:* the height of the valve on a liquid storage tank from which liquid exits the tank.

drip pot *n:* a device installed in a flow recorder's manifold to collect liquid that may condense out the gas in the manifold and to minimize the chance of inaccurate differential and static readings caused by liquid in the manifold lines.

drip time *n:* the amount of time it takes for the liquid clinging to the sides of a tank prover to fall and drip out of the vessel.

dry-bottom prover *n:* a type of open tank prover that has a gauge glass on the neck of the prover but not on the bottom of the prover. Compare *wet-bottom prover.*

dual-chamber fitting *n:* an orifice meter fitting having two chambers. The orifice plate is rolled from one chamber to another; the lower chamber is isolated from the upper chamber so that the gas flow can continue unimpeded while the plate is removed. Compare *single-chamber fitting.*

effective inside tank height *n:* the distance from the gauge point on the tank floor or datum plate to the top of the top angle or where the tank contents would begin to overflow.

elbow *n:* a fitting that allows two pipes to be joined at an angle of less than 180°, usually 90° or 45°.

emulsion *n:* a mixture in which one liquid, termed the dispersed phase, is uniformly distributed (usually as minute globules) in another liquid, called the continuous phase or dispersion medium. In an oil-in-water emulsion, the oil is the dispersed phase and the water the dispersion medium; in a water-in-oil emulsion, the reverse holds.

emulsion breaker *n:* see *demulsifier.*

entrained *adj:* drawn in and transported by the flow of a fluid.

equilibrium *n:* a state of balance between opposing forces or actions that is either static or dynamic.

expansion factor *n:* a multiplying factor used when calculating gas flow rate. This factor corrects for the reduction in fluid density that a compressible fluid experiences when it passes through an orifice as a result of the increased fluid velocity and the decreased static pressure.

expansion loop *n:* a loop built into a pipeline to allow for expansion and contraction of the line.

extractor *n:* in automatic sampling, a device for removing small amounts of liquid from a flowing stream and diverting these to a storage container.

fitting *n:* a small, often standardized, part (such as a coupling, valve, or gauge) installed in a larger apparatus.

flanged fitting *n:* a two-piece orifice fitting with flanged faces that are bolted together.

floating tap system *n:* a type of tap sampling system found on liquid storage tanks. The permanently installed taps on the outside of the tank are attached to pipes and a floating boom inside the tank that rises and falls with the liquid level.

flow monitor *n:* any device that senses the state of fluid flowing in a pipeline or other container.

flow-proportional sample *n:* a sample taken from a pipeline during the entire period of transfer of a batch at a rate that is proportional to the rate of flow of the liquid in the pipeline at any instant.

flow rate *n:* the speed, or velocity, of fluid flow through a pipe or vessel.

flow-sensing device *n:* a tool inserted into a pipeline or other container that can sense the flow of fluid within the container.

flow velocity *n:* the speed at which a fluid flows through a pipeline or other container.

free liquid *n:* hydrocarbons that exist as liquids and fall out of gaseous hydrocarbons.

frictional resistance *n:* the resistance to flow a fluid experiences when it is flowing through a pipeline or other container. Friction occurs between the fluid and the container's walls and within the fluid itself. Frictional resistance robs a flowing fluid of its pressure.

fuel oil *n:* a petroleum product, such as diesel fuel, that has a relatively low volatility.

gasoline *n:* a volatile, flammable liquid hydrocarbon refined from crude oils and used universally as a fuel for internal-combustion, spark-ignition engines.

gauge *n:* a device (such as a pressure gauge) used to measure some physical property. *v:* to measure size, volume, depth, or other measurable property.

gauge glass *n:* a glass tube or metal housing with a glass window that is connected to a vessel to indicate the level of the liquid contents.

gauge hatch *n:* the lidded opening in the top of an oil or oil product storage tank through which gauging and sampling operations are carried out.

gauge height *n:* see *reference height.*

gauge lines *n pl:* small-diameter pipes leading from an orifice fitting to a bellows in a flow recorder. They allow gas pressure to be exerted on the bellows so that the pressure drop across the orifice plate can be recorded.

gauge path *n:* the vertical distance from the reference point on the gauge hatch to the bottom of the tank or to the datum plate.

gauge pipe *n:* a vertical pipe that extends from the gauge hatch to the bottom of the tank or to the datum plate.

gauge point *n:* 1. the point at which a tape is lowered and read on a tank, usually at the rim of the hatch, manway, or expansion dome. 2. a point to which all subsequent measurements are related. 3. the point from which the reference height is determined and from which the ullages/innages are taken. Also called reference point, zero point.

gauger *n:* a pipeline representative who witnesses the sale or transfer of crude oil from the producer to the pipeline. This person samples or tests the crude oil to determine quantity and quality and uses a calibrated, flexible-steel tape with a plumb bob at the end to measure the oil in the tank.

gauger's bob *n:* a graduated weight used with a graduated tape to measure the height of the liquid in a storage tank. The weight of the bob keeps the tape taut in order to ensure a correct reading of the liquid level.

gauger's tape *n:* a graduated, metal, nonsparking, noncorrosive measuring tape used with a gauger's bob to measure the height of the liquid in a storage tank.

gauging height *n:* the distance from the gauge, or reference, point from the bottom of a tank to a gauge, or reference, point at the top of the tank, usually on the hatch. On marine vessels, the measurement must be made when the vessel is on even keel.

gear pump *n:* a rotary pump that moves fluid by means of two contrarotating, meshing gear wheels. The teeth of one wheel mesh with the spaces between the teeth of the other wheel. The fluid enters one side of the pump, is forced through by the moving gears, and is discharged on the other side.

grab *n:* in sampling petroleum and petroleum products, a small sample, usually 1.5 millilitres, obtained by the probe and sampler in an automatic sampling system. Several grabs constitute a sample.

gravity *n:* see *density.*

grind-out machine *n:* see *centrifuge.*

gross volume *n:* the total amount of liquid in a storage tank excluding any adjustments for S&W, temperature, or density. Compare *net volume.*

hatch *n:* 1. an opening in the roof of a tank through which a gauging line may be lowered to measure its contents. 2. the opening from the dock into the cargo space of ships.

hydrocarbons *n pl:* organic compounds of hydrogen and carbon whose densities, boiling points, and freezing points increase as their molecular weights increase. Although composed of only two elements, hydrocarbons exist in a variety of compounds, because of the strong affinity of the carbon atom for other atoms and for itself. The smallest molecules of hydrocarbons are gaseous; the largest are solids. Petroleum is a mixture of many different hydrocarbons.

hydrometer *n:* an instrument with a graduated stem, used to determine the density of liquids. The liquid to be measured is placed in a cylinder, and the hydrometer placed in the liquid. The hydrometer floats at a certain level in the liquid, depending on the density of the liquid.

ID plate *n:* on a flow recorder, the record of the static and differential spring ranges.

impeller *n:* a set of mounted blades used to impart motion to a fluid (e.g., the rotor of a centrifugal pump).

inferential meter *n:* a volume meter with the addition of a densitometer from which mass flow is inferred.

in-line system *n:* a type of automatic sampling system where the probe and extractor are located directly on the pipeline and the sample is withdrawn from the liquid stream. Compare *sample loop* .

innage bob *n:* a pointed, cylindrical weight attached to the end of a gauge tape. The tape and bob are used to measure liquid height in a tank. The bob keeps the tape vertical so that an accurate reading can be obtained from the tape.

innage measurement *n:* a measure of the liquid height in a tank from the bottom of the tank or datum plate to the top of the liquid.

innage tape-and-bob procedure *n:* a method of measuring the height of the liquid in a tank using a pointed gauger's bob and a graduated tape. In this method, the liquid height is measured from the bottom of the liquid to the top. Compare *outage tape-and-bob procedure.*

insulator *n:* a material with a very low concentration of free electrons that resists the flow of electric current.

kerosene *n:* a light, flammable hydrocarbon fuel or solvent.

LACT *abbr:* lease automatic custody transfer.

LACT unit *n:* an automated system for measuring, sampling, and transferring oil from a lease gathering system into a pipeline. See also *lease automatic custody transfer.*

lease *n:* 1. the area where production wells, stock tanks, separators, LACT units, and other production equipment are located. 2. a legal document executed between a landowner, as lessor, and a com-pany or individual, as lessee, that grants the right to exploit the premises for minerals or other products; the instrument that creates a leasehold or working interest in minerals.

lease automatic custody transfer (LACT) *n:* the measurement, sampling, and transfer of oil from the producer's tanks to the connected pipeline on an automatic basis without a representative of either the producer or the gathering company having to be present. See also *LACT unit.*

lease operator *n:* the oil company employee who attends to producing wells. This per-son attends to any number of wells, ensures steady production, prepares reports, tests, gauges, and so forth. Also called a custodian, pumper, or switcher.

light ends *n pl:* the lighter hydrocarbon molecules that comprise gasoline, light kerosene, heptane, natural gas, and so forth.

light fuels *n pl:* fuels, such as gasoline, that have relatively high volatility.

light products *n pl:* petroleum fractions, such as ethylene, that are relatively light in molecular weight.

linear chart *n:* a direct reading chart used on the flow recorder of an orifice metering system on which are recorded static and differential pressures during a particular time period. It must have the same static and differential ranges as the flow recorder.

liquid calibration procedure *n:* a method of determining a tank's capacity by filling and withdrawing from the tank accurately determined volumes of liquid.

liquid cut *n:* the mark on the gauger's tape indicating the height of the liquid.

liquid displacement *n:* the crude oil or product that is forced into the hatch or up around the edges of a floating roof by the weight of the roof.

liquid head stress *n:* the hydrostatic pressure of the liquid in tanks of 500 barrels or more, which causes the shell to expand and contract.

liquid-in-glass thermometer *n:* a type of thermometer in which the sensitive part of the instrument consists of a liquid like mer-cury or alcohol contained in an envelope of glass.

lobed impeller *n:* a rounded blade on a dis-placement meter that separates the liquid stream into discrete quantities.

logarithmic scale *n:* a scale in which the distances of numbers from a reference point are proportional to the logarithms of the numbers.

L-10 chart *n:* a chart used in measuring gas with orifice meters on which are recorded differential and static pressures. It uses a logarithmic scale from 0 to 10. The readings must be converted, but it can be used on any flow recorder, regardless of range. Also called square-root chart.

manifold valve *n:* a valve placed in a series of piping that is called a manifold.

manometer *n:* a U-shaped piece of glass tubing containing a liquid (usually water or mercury) that is used to measure the pressure of gases or liquids. When pressure is applied, the liquid level in one arm rises while the level in the other drops. A set of calibrated markings beside one of the arms permits a pressure reading to be taken, usually in inches or millimetres.

mass flow rate *n:* a calculation of gas flow through an orifice meter where the quantity of the gas (independent of external conditions like temperature, pressures, or gravity) is determined; usually expressed in pounds mass per unit time.

master meter *n:* a meter that is proved using a certified prover and then used to calibrate other provers or to prove other meters.

master tape *n:* a measuring tape that has been certified accurate at a specified temperature and tension by the NIST.

measuring chamber *n:* the portion of a displacement meter that contains the measuring element.

meniscus *n:* the curved upper surface of a liquid column, concave when the containing walls are wet by the liquid (negative meniscus) and convex when not (positive meniscus).

meter *n:* a device used to measure and often record volumes, quantities, or flow rates of gases, liquids, or electric currents. *v:* to measure quantities or properties of a substance.

meter chart *n:* in gas measurement, a circular chart of special paper that shows the range of differential pressure and static pressure and that is marked by the recording pens of a flow recorder in an orifice metering system.

meter factor *n:* a number used to correct a meter's inaccuracy. The factor is derived by dividing the actual volume of liquid passed through a meter during proving by the volume registered by the meter. For subsequent metering operations, the actual throughput, or gross measured volume, is determined by multiplying the indicated volume registered on the meter by the meter factor.

metering manifold *n:* a collection of pipe, a manifold, in which is mounted a meter for measuring fluid flow through the manifold. Usually a meter is mounted in a manifold so fluid flow is diverted into it for measurement.

meter-line thermometer *n:* a thermometer located in a pipeline either immediately upstream or immediately downstream of the meter; used to determine the temperature of the liquid flowing through the meter.

meter run *n:* see *orifice meter installation.*

meter skid *n:* a small platform fabricated from pipe or bar stock and steel plate onto which a meter and manifold is mounted.

meter station *n:* see *orifice meter installation.*

meter tube *n:* an important part of the primary element of an orifice meter installation that must create a known flow pattern for the fluid as it reaches the plate. It is the straight upstream pipe of the same size be-tween the orifice and the nearest pipe fitting and the similar downstream pipe between the orifice and the nearest pipe fitting.

micrometer *n:* 1. a caliper for making precise measurements; a spindle is moved by a screw thread so that it touches the object to be measured. 2. an instrument used with a telescope or microscope to measure minute distances.

monitor *n:* an instrument that reports the performance of a control device or signals if unusual conditions appear in a system. For example, an S&W monitor provides a mechanical means of preventing contaminated oil from entering the pipeline by detecting the presence of excessive water and actuating valves to divert the flow back to dehydration facilities.

naphtha *n:* a volatile, flammable liquid hydrocarbon distilled from petroleum and used as a solvent or a fuel.

natural gas *n:* a highly compressible, highly expansible mixture of hydrocarbons with a low specific gravity and occurring naturally in a gaseous form. Besides hydrocarbon gases, natural gas may contain appreciable quantities of nitrogen, helium, carbon dioxide, hydrogen sulfide, and water vapor. Although gaseous at normal temperatures and pressures, the gases making up the mixture that is natural gas are variable in form and may be found either as gases or as liquids under suitable conditions of temperature and pressure.

net volume *n:* the total volume of liquid in a tank after adjustments have been included for S&W content, temperature, and density. Compare *gross volume.*

NIST *abbr:* the National Institute of Standards and Technology.

nominal volume *n:* the quantity assigned to a tank or vessel for the purpose of identification only; the exact volume may be somewhat different from the nominal volume.

observed volumes *n pl:* those volumes of liquid or gas that are indicated by the register on the meter. Compare *true volume.*

octane level *n:* a measure of the antiknock properties of a fuel.

oil-water interface *n:* the boundary between the oil on top and the water below when both are contained in the same vessel.

opaque liquid *n:* a liquid, such as crude oil, that is dark and difficult to see through; light does not penetrate it.

opening gauge *n:* the measurement of liquid in a tank before a delivery or receipt. Compare *closing gauge.*

open-tank prover *n:* a relatively small calibrated tank, open to the atmosphere. Often used to prove meters at loading racks. In use, the meter to be proved is set at zero and the prover tank is filled to a calibrated height while the meter registers the amount of liquid flowing through it. When the tank is filled, the calibrated volume in it is compared to the volume the meter registered. If the meter is inaccurate, it is adjusted or a factor is used to compensate for the inaccuracy.

operator *n:* see *producer.*

optical scanner *n:* a device in which a moving spot of light controlled mechanically or electronically scans the meter chart. The reflected light generates signals that are used to process information from the chart.

orifice *n:* an opening of a measured diameter that is used for measuring the flow of fluid through a pipe or for delivering a given amount of fluid through a fuel nozzle. In measuring the flow of fluid through a pipe, the orifice must be of smaller diameter than the pipe diameter. It is drilled into an orifice plate held by an orifice fitting.

orifice meter installation *n:* in gas measurement, the orifice plate, orifice fitting, manifold, and flow recorder. Also called meter run, meter station.

orifice plate *n:* a sheet of metal, usually circular, in which a hole of specific size is made for use in an orifice fitting.

outage bob *n:* a graduated plumb bob that is attached to a metal gauge tape and that is of sufficient weight to keep the tape taut.

outage measurement *n:* a measure of the space between the top of the liquid in a tank and the reference point on the hatch. The liquid level in the tank is determined by subtracting the outage measurement from the gauging height of the tank.

outage tape-and-bob procedure *n:* a method of measuring the liquid level in a tank by measuring the space between the top of the liquid in a tank and the reference point on the hatch. The liquid level in the tank is determined by subtracting the outage measurement from the gauging height of the tank. Compare *innage tape-and-bob procedure.*

parameters *n pl:* the values that characterize and summarize the essential features of measurements.

percentage chart *n:* a chart for reading differential and static pressures. Its readings reflect a percentage of full scale from 0 to 100 for both static and differential pressure. Readings must be converted, but can be used on any flow recorder, regardless of range.

pipe prover *n:* 1. an accurately calibrated, usually U-shaped, pipe used to check, or prove, the accuracy of a meter on a LACT unit or other oil measuring installation. 2. a type of continuous-flow volumetric prover comprising a length of pipe from which a known volume is displaced by a displacer to or from a meter being proved at normal operating conditions.

pop valve *n:* see *pressure relief valve.*

posted gauge height *n:* the vertical distance from the bottom of the tank or datum plate to the reference point on the hatch; normally engraved on or near the reference point.

power mixers *n pl:* in a sample container, a device that is moved by an electric motor to thoroughly mix the liquid in the container.

preheater *n:* a heater used to warm the sample before and during centrifuging to determine suspended S&W content.

pressure base factor *n:* a factor used in the formula recommended prior to August 1992 for the calculation of gas volume flow. The factor adjusted the volume to a standard atmospheric pressure regardless of the location.

pressure differential *n:* see *differential pressure.*

pressure gauge *n:* an instrument that measures fluid pressure and usually registers the difference between atmospheric pressure and the pressure of the fluid by indicating the effect of such pressures on a measuring element (e.g., a column of liquid, pressure in a Bourdon tube, a weighted piston, or a diaphragm).

pressure-reducing valve *n:* a type of valve that changes the upstream pressure to a lower downstream pressure.

pressure relief valve *n:* a valve that opens at a preset pressure to relieve excessive pressures within a vessel or line. Also called a pop valve, relief valve, safety valve, or safety relief valve.

pressure surge *n:* a sudden increase in pressure in fluid flowing in a pipe or in liquid flowing into a storage tank.

pressure tap *n:* in an orifice fitting, the threaded holes on each side of the orifice plate. Small pipes are screwed into the holes to connect the fitting with a flow recorder. The taps are used so that pressure differential on either side of an orifice plate can be recorded.

primary element *n:* that part of an orifice meter installation that creates the pressure drop in the pipeline necessary for gas measurement; includes the meter tube, orifice plate, fitting, and the pressure taps.

probe *n:* any small device that, when brought into contact with or inserted into a system, can make measurements on that system. In sampling, probes are inserted into pipelines so that small samples of liquid can be removed and then stored for future testing.

producer *n:* 1. a well that produces oil or gas in commercial quantities. 2. an operating company or individual in the business of producing oil; commonly called the operator.

prove *v:* to determine the accuracy of a petroleum measurement meter.

prover *n:* a device used to determine the accuracy of a petroleum measurement meter.

prover connector valve *n:* a device used to connect the prover to the pipeline either upstream or downstream of the meter to be proved.

psia *abbr:* pounds per square inch absolute.

psig *abbr:* pounds per square inch gauge.

pulsating flow *n:* flow that is variable; unstable flow.

pulse generator *n:* 1. a generator that produces repeating pulses or pulses initiated by signals. 2. a displacement meter accessory coupled to the measuring element and designed to produce a series of electrical pulses whose number is proportional to the volume measured and whose frequency is proportional to the flow rate. Also called pulser.

pulse monitor *n:* a device in a pipe prover system that senses and counts the number of pulses put out by a pulse generataor on the meter being proved. Pulse count relates to the amount of fluid that flows through the meter.

pulser *n:* see *pulse generator.*

pumper *n:* see *lease operator.*

Rankine temperature scale *n:* a temperature scale with the zero point at absolute zero. On the Rankine scale, water freezes at 491.60° and boils at 671.69°.

receiver *n:* a pressure-tight storage container for liquid samples.

reference height *n:* the distance from the reference point to the datum plate or the bottom of the tank. It should be stamped on the fixed benchmark plate or stenciled on the tank roof near the gauging hatch. Also called gauge height.

reference point *n:* 1. the point at which a tape is lowered and read on a tank, usually at the rim of the hatch, manway, or expansion dome. 2. a point to which all subsequent measurements are related. 3. the point from which the reference height is determined and from which the ullages/innages are taken. Also called gauge point, zero point.

registered volume *n:* the amount of a substance as indicated by the register on the meter.

Reid vapor pressure *n:* the vapor pressure of a liquid at 100° Fahrenheit (37.78° Celsius, 311° Kelvin) as determined by ASTM D 323-58, Standard Method of Test for Vapor Pressure of Petroleum Products (Reid Method).

relative density *n:* 1. the ratio of the weight of a given volume of a substance at a given temperature to the weight of an equal volume of a standard substance at the same temperature. For example, if 1 cubic inch of water at 39°F weighs 1 unit and 1 cubic inch of another solid or liquid at 39°F weighs 0.95 unit, then the relative density of the substance is 0.95. In determining the relative density of gases, the comparison is made with the standard of air or hydrogen. 2. the ratio of the mass of a given volume of a substance to the mass of a like volume of a standard substance, such as water or air. Also called specific gravity.

relief valve *n:* see *pressure relief valve.*

representative sample *n:* a small portion extracted from the total volume of material that contains the same proportions of the various flowing constituents as the total volume of liquid being transferred. The precision of extraction must be equal to or better than the method used to analyze the sample.

rosin *n:* a translucent amber-colored to almost black brittle friable resin obtained by chemical means from the oleoresin or dead wood of pine trees or from tall oil and used especially in making varnish, paper size, soap, and soldering flux and in rosining violin bows.

rotor *n:* 1. a device with vanes or blades attached to a shaft. The device turns or rotates when the vanes are struck by a fluid directed there by a stator. 2. the rotating part of an induction-type alternating current electric motor.

running sample *n:* in tank sampling, a sample obtained by lowering an unstoppered beaker or bottle from the top of the oil to the level of the bottom of the outlet connection or swing line and returning it to the top of the oil at a uniform rate of speed so that the beaker or bottle is about three-quarters full when withdrawn from the oil.

safety relief valve *n:* see *pressure relief valve.*

safety valve *n:* see *pressure relief valve.*

sample-loop *n:* a relatively small pipe that is a bypass from the main pipeline. Liquid in the pipeline also flows through the loop, from which the liquid can be sampled.

S&W *abbr:* sediment and water. API Committee on Petroleum Measurement prefers this abbreviation to the older "BS&W."

S&W probe *n:* a small device inserted in a pipeline to measure the amount of S&W in the liquid flowing through the line.

secondary element *n:* that part of an orifice meter installation that contains the recording elements, including the gauge lines, orifice meter chart, flow recorder, clock, static spring, etc.

sediment *n:* 1. the matter that settles to the bottom of a liquid; also called tank bottoms. 2. in geology, buried layers of sedimentary rocks.

sediment and water (S&W) *n:* a material coexisting with, yet foreign to, petroleum liquid and requiring a separate measurement for reasons that include sales accounting. This foreign material includes free water and sediment (dynamic measurement) and/or emulsified or suspended water and sediment (static measurement). The quantity of suspended material present is determined by a centrifuge or laboratory testing of a sample of petroleum liquid.

semiliquid *n:* a substance that is intermediate between a liquid and a solid.

sensing element *n:* the part of a measuring device that is responsive to the absolute value or change in a physical quantity, such as temperature, pressure flow rate, pH, light, or sound. It converts that change into an input signal for an information-gathering system.

shake-out machine *n:* see *centrifuge.*

shell height *n:* the distance between the bottom of the bottom angle of the tank and the top of the top angle.

single-chamber fitting *n:* in an orifice meter installation, a mechanism for holding the orifice plate in the pipeline. In a single-chamber fitting, the gas flow must be stopped and the meter tube depressured before the orifice plate can be removed. Compare *dual-chamber fitting.*

single-chamber fitting with gearing *n:* in an orifice meter installation, a type of single-chamber fitting that has a lever and gears that can be used to roll the orifice plate up and out of the fitting.

slippage *n:* liquid that slips between the blade or lobed impeller of a displacement meter and the meter housing instead of being captured between the two blades or impellers and being measured.

solubility *n:* the capacity of a substance to be dissolved in another.

solvent *n:* a substance, usually liquid, in which another substance (the solute) dissolves.

specific gravity *n:* 1. the ratio of the weight of a given volume of a substance at a given temperature to the weight of an equal volume of a standard substance at the same temperature. For example, if 1 cubic inch of water at 39°F weighs 1 unit and 1 cubic inch of another solid or liquid at 39°F weighs 0.95 unit, then the relative density of the substance is 0.95. In determining the relative density of gases, the comparison is made with the standard of air or hydrogen. 2. the ratio of the mass of a given volume of a substance to the mass of a like volume of a standard substance, such as water or air. Also called relative density.

spirit level *n:* a device consisting of a bubble tube designed to determine a horizontal line or plane.

spot sample *n:* in tank sampling, a sample obtained at some specific location in the tank by means of a thief, a bottle, or a beaker.

square-root chart *n:* see *L-10 chart.*

standard tape *n:* the measuring tape used to measure the circumference of the ring at the bottom of a tank. It is calibrated by the National Institute of Standards and Technology and is usually 100 feet long.

static mixer *n:* a mixer with no moving parts. Kinetic energy of the moving fluid provides the power for mixing. An example is a jet hopper.

static pressure *n:* the pressure exerted by a fluid on a surface that is at rest in relation to the fluid.

static range *n:* the extent of variation between the lowest and highest static pressures that a particular orifice meter can record.

static spring *n:* in gas measurement by orifice meters, the spring in a flow recorder whose size and strength determines the range over which static pressure will be recorded.

strap *v:* to measure and record the dimensions of oil tanks to prepare tank tables (gauging tables) for determining accurately the volume of oil in a tank at any measured depth.

strapping pole *n:* an aluminum pole of adjustable length used on several rings of a tank while marking the working tape's path around the tank.

stress a tank *v:* to fill a tank completely with a liquid that has the same or greater density as the liquid that will normally be in the tank.

striking point *n:* a spot on the bottom of a storage tank or on the datum plate that is directly below the reference point on the hatch. This location is where the innage bob comes to rest when the tank is gauged and serves as the zero point for all innage measurements.

sulfur (S) *n:* a pale yellow, nonmetallic chemical element. In its elemental state—free sulfur—it has a crystalline or amorphous form. In many gas streams, sulfur may be found as volatile sulfur compounds—hydrogen sulfide, sulfur oxides, mercaptans, carbonyl sulfide. Reduction of their concentration levels is necessary for corrosion control and, in many cases, necessary for health and safety reasons.

suspended S&W *n:* sediment and water that are suspended in oil and that can be separated only by (1) centrifuge with appropriate solvents or (2) extraction by distillation.

switcher *n:* see *lease operator.*

tank bottoms *n pl:* the settlings in the bottom of a storage tank.

tank calibration *n:* see *tank strapping.*

tank strapping *n:* measuring a tank in order to obtain certain of its dimensions, such as the depth of the tank inside and outside, the circumference of each ring on the tank, and the height of the liquid in the tank.

tape clamp *n:* a quick-release clamp that may be fitted around a strapping tape at any convenient position throughout its length.

tee *n:* a pipe fiting that is shaped like the letter T. A pipe or other fitting can be attached to each end of the tee, since the tee ends are threaded.

temperature compensator *n:* a device used on a meter automatically to convert the flowing temperature of the fluid through the meter to a standard temperature.

tender number *n:* a number assigned to a particular shipment, or tender, of petroleum or petroleum product.

tension gauge *n:* a gauge that tells a strapping crew how much tension is being applied to a measuring tape.

tetraethyl lead *n:* an antiknock compound for motor gasoline. It is being phased out because lead is a pollutant.

thermohydrometer *n:* a combination hydrometer-thermometer designed to measure density and temperature of a liquid.

thermometer *n:* an instrument that measures temperature. Thermometers provide a way to estimate temperature from its effect on a substance with known characteristics (such as a gas that expands when heated). Various types of thermometers measure temperature by measuring the change in pressure of a gas kept at a constant volume, the change in electrical resistance of metals, or the galvanic effect of dissimilar metals in contact. The most common thermometer is the mercury-filled glass tube, which indicates temperature by the expansion of the liquid mercury.

thief *n:* a device that is lowered into a tank to take an oil sample at any depth. The sample will subsequently be used to deter-mine the quality and S&W content of the oil in the tank. *v:* to obtain a sample of oil from a tank using a thief.

threaded coupling *n:* a type of connector that has threads on each end, making it pos-sible to screw two pieces of pipe together.

throughput volume *n:* the volume of fluid that has flowed through a meter during a specific amount of time.

time-proportional sample *n:* an automatic sample usually obtained from a pipeline and taken at regular timed intervals.

tolerance *n:* the range of variation permitted in maintaining a specific dimension in machining a part. For example, a shaft measurement could be 2.000 inches ± 0.001 (50 millimetres ± 0.025). This means that any measurement from 1.999 inches to 2.001 inches (49.975 millimetres to 50.025 millimetres) would be acceptable.

top angle *n:* the angle on a storage tank where the side wall is joined to the roof of the tank; it is usually 90°.

transducer *n:* a device actuated by power from one system and supplying power to another system, usually in a different form. For example, a telephone receiver receives electric power and supplies acoustic power. In orifice meter installation, the transducer receives the differential or static pressure and supplies electronic power to a computer.

trip rod *n:* the stem on the end of a core thief that can be bumped or tapped to close and thus catch a sample in the thief.

true volume *n:* the volume of fluid flowing through a meter, including adjustments for meter factor, temperature, density, pressure, etc. Compare *observed volumes.*

trunnion cup *n:* in a centrifuge, the container into which a sample is placed for centrifuging.

T square *n:* a ruler with a crosspiece or head at one end used in making parallel lines.

turbine fuel *n:* a fuel oil, similar to diesel fuel, which is used to run gas turbines.

turbine meter *n:* a velocity-measuring device for fluids in which the flow is parallel to the rotor axis and the speed of rotation is proportional to the rate of flow. The volume of gas in gas measurement is determined by counting the revolutions of the rotor. In liquid turbine meter measurement, the meter and electronic instrumentation are combined to measure total flow and/or flow rate within the piping system.

turbulent flow *n:* the erratic, nonlinear flow of a fluid, caused by high velocity. Characterized by random eddying flow patterns superimposed on the general flow progressing in a given direction.

ullage tape-and-bob procedure *n:* see *outage tape-and-bob procedure..*

U-tube *n:* a U-shaped tube. *v:* to cause fluids to flow in a U-tube without adding an outside force.

vacuum breaker *n:* a device used to prevent a vacuum from occurring in a tank, vessel, piping, and so on.

valve *n:* a device used to control the rate of flow in a line to open or shut off a line completely, or to serve as an automatic or semiautomatic safety device. Those used extensively include the check valve, gate valve, globe valve, needle valve, plug valve, and pressure relief valve.

vapor *n:* a substance in the gaseous state that can be liquefied by compression or cooling.

vapor loss *n:* the loss of hydrocarbons through evaporation.

vapor pressure *n:* the pressure exerted by the vapor of a substance when the substance and its vapor are in equilibrium. Equilibrium is established when the rate of evaporation of a substance is equal to the rate of condensation of its vapor.

vapor recovery unit (VRU) *n:* 1. in production operations, equipment designed to recover light ends that are released from oil in a stock or other tank. 2. in petroleum refining, a process unit consisting of a scrubber and a compressor. It is designed to recover petroleum from hydrocarbon vapors and safely to handle toxic gases produced from some wells.

velocity meter *n:* a type of meter that measures the rate of fluid flow through it in order to determine the quantity of fluid flowing through it.

vena contracta *n:* squeezing of the gas flow.

vertical loop *n:* a change in the configuration of a pipeline whereby the pipe is curved from the horizontal to a 90° vertical run, then curved 90° to a horizontal run, curved again to a 90° vertical, and then curved 90° back to horizontal. The effect is to create turbulence in the fluid flow and distribute the S&W evenly in the flowing stream.

viscosity *n:* a measure of the resistance of a fluid to flow. Resistance is brought about by the internal friction resulting from the combined effects of cohesion and adhesion. The viscosity of petroleum products is commonly expressed in terms of the time required for a specific volume of the liquid to flow through a capillary tube of a specific size at a given temperature.

viscous *adj:* having a high resistance to flow.

volatile *adj:* readily vaporized.

volatility *n:* the tendency of a liquid to assume the gaseous state.

volume flow rate *n:* the amount, or quantity, of a fluid that passes a point for a given time.

water cut *n:* the percentage of water in fluid produced from a well. *v:* to locate the oil-water interface and use the location to measure the volumes of oil and water in a shore tank or ship compartment.

water saturation *n:* the percentage of total pore space in a reservoir occupied by water. It is normally estimated by electric log calculations, although cores specially obtained may be analyzed for this value.

weir *n:* a metal plate installed in a separator, treater, or mud tank and used to regulate liquid level or, in the case of a metering separator, to measure flow.

wet-bottom prover *n:* an open tank prover with a gauge glass on the bottom of the prover. Compare *dry-bottom prover.*

wide-line chart *n:* an orifice chart where the differential and static pressure recordings indicate a wide range of pressure readings.

working tape *n:* the measuring tape used to measure all rings after the ring at the bottom of the tank. It is calibrated to the standard tape.

zero *v:* to bring a differential pen to the zero point on an orifice chart.

zero point *n:* see *gauge point.*

Index

D

E

F

G

H

I

L

M

N

O

T